Building Partnerships

How to Work With Trade Contractors

BOB WHITTEN

National Association of Home Builders
1201 15th Street, NW
Washington, DC 20005-2800
(800) 223-2665
www.BuilderBooks.com

Building Partnerships
How to Work With Trade Contractors

BuilderBooks, a Service of the National Association of Home Builders

Christine B. Charlip	Publisher
Doris M. Tennyson	Senior Acquisition Editor
Torrie Singletary	Production Editor
David Rhodes	Cover Design
Harlowe	Composition
McNaughton & Gunn, Inc.	Printing
Gerald M. Howard	NAHB Executive Vice President and CEO
Mark Pursell	NAHB Senior Staff Vice President, Marketing & Sales Group
Lakisha Campbell	NAHB Staff Vice President, Publication & Affinity Programs

This publication is designed to provide accurate and authoritative information in regard to the subject matter covered. It is sold with the understanding that the publisher is not engaged in rendering legal, accounting, or other professional service. If legal advice or other expert assistance is required, the services of a competent professional should be sought.

—From a Declaration of Principles jointly adopted
by a Committee of the American Bar Association
and a Committee of Publishers and Associations

Printed in the United States of America

10 09 08 07 06   1 2 3 4 5

Library of Congress Cataloging in Publication Data

Whitten, Bob, 1950-
Building Partnerships : how to work with trade contractors
/ Bob Whitten
p.   cm.
ISBN-10: 0-86718-483-3
ISBN-13: 978-0-86718-483-9
1. Construction industry—Management.   2. Building
trades—Management.   3. Contractors.   I. Title.
HD9715.A2 W468 1999
624'.068—dc21

99-41436
CIP

For further information, please contact:

1201 15th Street, NW
Washington, DC 20005-2800
800-223-2665
Visit us online at www.BuilderBooks.com.

Table of Contents

Acknowledgments

The author and BuilderBooks wish to acknowledge the contributions of the following reviewers who assisted with reviews of the proposal or the developing manuscript: Bill Adams, Jerry Householder, members of The Johnson Management Group, Leon Rodgers, Fred White, Stephen K. Hann, Jim Billman, Manuel Dembs, Kim Post, Dottie Piazza, Alan Simonini, Theodore Visnic, Steve Swapford, Tom Burns, Fred Parker, members of the F.A.S.T. Management Group, and Bill Eich.

The author also wishes to thank The F.A.S.T. Management Group of Redmond, Washington and Leon Rogers of Brigham Young University for sharing their scheduling success formulas and acknowledges Fred White and Dave Showers, former mentors at Wayne Homes, and Dan Cooper of Cooper Communities for allowing me the opportunity to work with NAHB on projects such as this. In addition, the author would like to thank the hundreds of trade contractors (in Ohio and Alabama) who have made valuable contributions to this book and to the industry at large in my home markets for the past 15 years.

Finally, the author wishes to thank his partners Randy Hymas and Russel Neumann for keeping the field construction going while this edition of the book was "under construction." Additionally, intense gratitude is offered to editor Kurt Lindblom for his understanding and patience during this revision.

About the Author

Bob Whitten is an Executive Vice President of Cooper Homes, Inc. (CHI). CHI is a wholly owned subsidiary of Cooper Communities, Inc. CHI is the largest builder in Arkansas with divisions in Bella Vista and Hot Springs, AR, Tellico Village, TN, and Savannah Lakes, SC. From 1997 to March of 1999, Mr. Whitten was managing partner of Vista South Custom Homes in Madison, AL. Vista South is a design-build firm specializing in innovative custom homes and is the newest division of CHI.

From 1989 to 1997, Mr. Whitten was Chief Operations and Financial Officer for Wayne Homes of Ohio. During the mid 1980s, Mr. Whitten was the Director of Business Management for the National Association of Home Builders, where he directed efforts to educate builders in areas of successful business management. He was the founding editor of NAHB's *Builder's Management Journal.*

Mr. Whitten has worked on several past editions of NAHB's *The Single-Family Builder Cost of Doing Business* study and was the principal author of the 1991 edition of that study. He has been a speaker at the NAHB annual convention each year for the past decade.

Introduction

Almost Eight Years Later

I t doesn't seem possible that almost eight years have passed since the first edition of this book. Much has happened. The industry has seen significant changes. I have worked in building companies that have built more than 3,500 homes since 1991. I have been forced into the day-to-day builder or general contractor and trade contractor relationship much more closely than when I concentrated my efforts on the financial side of our building operations. I think that has been a good thing.

Over the past two years, I made the transition from an officer of a large building company in the great state of Ohio to a principal partner in a small custom building company in northern Alabama. I have been exposed to trade contractors of all shapes, sizes, and backgrounds.

The home building and light commercial construction industry is facing an intense labor shortage in most U.S. markets. This has caused a reconsideration of the trade contractor/builder relationship. We will discuss this relationship in-depth in the first and again in the final chapters of the book. In short, however, the labor shortage has caused more of our industry problems in the areas of cost control, schedule overruns, and long punch lists than yesteryear.

This labor shortage and its effects caused my sense of humor to improve because if you take this business too seriously you'll go "nuts" as well as broke. You have to appreciate the differences in people, be flexible in approach, and learn that sometimes you have to compromise speed for quality, price for speed, or either for someone who knows the purpose of a dumpster.

On a serious note, I don't think our industry has reacted with enough foresight in facing this labor dilemma. In the future, we will need to find new production methods to meet the new construction demands. Controlled environment construction (a fancy term for building in a factory) will become a more practiced and cost effective way of life for all but the highest end of residential contractors during the next generation. Component systems will replace field craftsmen. The controlled environment approach will allow labor to be trained in smaller increments of any one trade and will allow wider cross training, etc.

This book will address some of these issues throughout the chapters. This new edition of the trade contractor book bears a new title and a new approach. The "guts" of the original manuscript remain, but the presentation has changed to provide more direct experience insight and to point out which techniques work best with smaller volume builders and which work best with production builders with multiple layers of management.

We will investigate some general "How to" topics in Chapter 1 "Working with Trade Contractors." Then in Chapters 2 through 6, we will investigate finding trades, contracts, pricing, quality, scheduling, and pitfalls to avoid. Customer relations will be the topic for Chapter 7, and in the concluding chapter, we will look into the magic ball to see if we can shed some intelligent light into the future of the builder and trade contractor relationship.

Appendix A expands the presentation of the trade contractor written agreements by including examples for a variety of trades. Appendix B includes comfort zone checklists for each of the major trade contractors used in residential and light commercial construction. Appendix C includes quality control checklists.

Let's get started, learn something perhaps, and have some fun.

1

Working with Trade Contractors

T rade contractors are an integral part of the residential and light commercial construction industry. Labor has become a key issue today. Inflation and dwindling labor supplies have forced wage rates up. It is not uncommon for labor to account for 40 to 45 percent of the direct construction cost and 30 to 33 percent of the sales price of a new home. Small increases in these percentages cause major increases in the price and affordability of new homes and commercial structures.

Price is not the only concern with labor. "Price, quality, and speed, take two" used to be a familiar adage. Today, it is hard to find one of these traits in a contracted labor force. Time after time, when asked to identify the biggest challenge they face on a day-to-day basis, construction managers respond with issues involving contracted labor.

The most frequently identified problems include finding enough trades, ensuring quality performance, lack of communication, and unprofessional attitudes and appearances. Another frequently mentioned problem is trades who require payment within hours of completion of their work. Some trades even try to get paid before the job is finished and act like everybody else in town does it, why won't you?

Construction managers also say they have problems getting trades to show up on schedule. An additional problem is dealing with trades whose idea of proper communication with customers and the general public is to tell them how unreliable the general contractor or builder is. Talk about biting the hand that feeds you!

This book is not going to make any of these issues disappear. My hope is to generate some creative thinking enabling you to avoid some of the consistent errors that have been made by general contractors and builders. I will provide you with some examples of what to do and not to do in certain situations and give some insights into systems, rules, and forms that builders have developed to minimize the labor-induced hair loss phenomena. And you thought it was the hard hats causing all of us to go bald!

What is a "Trade" Contractor?

One of the many changes in our industry in the last eight years is the need for political correctness. Today, what used to be called a "subcontractor" is now referred to as a "trade contractor." The popular thought being that the term "sub" somehow meant inferiority to the general contractor or builder.

Legal Definition of a Trade Contractor Relationship

Most of you reading this book understand the definition of a trade contractor. The Internal Revenue Service has gone to great length and expense over the past decade to tighten the regulations on, and definition of, a contractor versus an employee. The IRS's goal is to prevent employers (general contractors or builders in our industry) from hiding regular employees behind trade contracting disguises. The IRS wants everyone to pay the proper payroll taxes (like a contractor wouldn't).

The IRS has established a set of rules to help field agents and employers to determine when someone you pay is a contractor and not an employee. The following is a subset of these rules:

1. Contractor relationship is rarely tested when the trade contractor is incorporated.

2. Contractors must do business for more than one employer or builder.

3. Contractors should work a schedule that they determine.

4. Contractors should not be paid on an hourly basis.

5. Contractors should maintain their own liability insurance and worker's compensation where and when required.

6. Contractors should have their own office, stationery, and business cards.

7. Contractors should provide their own tools and equipment.

8. Contractors should endorse their checks with a rubber stamp versus their own personal signature. One of the main tests IRS field agents use to determine a contractor relationship is to pull a few canceled checks to determine how the contractor endorsed them. If they are stamped (especially if they are stamped with a company or D.B.A. name), the agents have been taught to assume that the contractor relationship is valid.

9. Trade contractors operating under a tax identification number different from their own social security numbers are less likely to have their contractor relationship challenged by the IRS.

Keep these suggestions and rules in mind to help ensure that the IRS recognizes your trade contractors as such. Consult with your tax and/or legal advisor for additional federal and state requirements as the laws regarding subcontracted labor are constantly being updated.

Why Not In-House Labor?

With all the difficulties mentioned thus far, why don't more builders use in-house labor or employees? The answer is specialization and the desire to establish a firm price for the work in question. Over the past few decades, builders have realized that trade specialists can do higher quality work, faster, and often for less cost than their own general labor crews. They have also realized that a fixed bid or contract price helps eliminate the uncertainty of how much a phase of construction will cost in terms of variable labor hours.

Today, most of the labor in U.S. residential construction is subcontracted. The use of this subcontracted labor seems to be a cornerstone of operational procedures of successful home builders. Some holdover framing and/or trim carpentry crews and general cleanup crews remain in some firms, but have become rare in the past few years.

Most of these firms have found supervision of internal labor crews an unbearable burden, especially, the problems associated with scheduling productive 40-hour (minimum) workweeks for each employee. Most builders add that, given the increasingly complex product that today's market demands, maintaining quality requires specialized labor. Specialized labor is too expensive for miscellaneous duties between trade assignments.

Using trade contracted labor can help builders more clearly ascertain their labor costs throughout their projects. A preconstruction labor contract, negotiated for a specific sum for a specific job performance, is much easier to budget than hourly wages with the associated taxes, worker's compensation, and fringe benefits. From a management standpoint, less supervision is generally required for contracted labor. Hourly labor sometimes encourages workers to fill the time available. Contracted labor makes more money by working faster and getting on to the next contract.

Clearly, most builders have good reason to use contracted trade labor. However, with contracted labor new methods of approaching the working relationship are necessary.

Partnering with Trades

The current term used to describe maintaining a good working relationship with your trade contractors is "partnering." Indeed, a successful general/trade contractor relationship is a partnership in profit. For these relations to succeed long term, both contractors must make a profit.

Partnering is actually an extension of the Total Quality Management (TQM) philosophy called "win-win." The win-win philosophy implies that the parties in a relationship (in this case the trade and the builder) understand each other's goals and work together to build a mutually profitable relationship. The relationship centers on an action plan that encourages each organization to succeed.

For decades, American businesspeople approached external business relationships with "dog-eat-dog" attitude. In the "dog-eat-dog" scenario, the trade is viewed as an opponent to be defeated at the negotiating table at every opportunity. Once under contract, the trade is used until the performance slips or the price rises—and then is quickly replaced. The trade works for the builder and likes it, or else. This one-sided approach is archaic and results in an eventual lack of trust between the parties.

If for no other reason, the labor shortage should be motivating builders and general contractors to use the win-win philosophy and partner with their trades. By partnering, builder and trade can each accomplish more. Equally important as the employer/employee relationship, the builder/trade relationship must be cultivated for everyone's benefit. Trades should be viewed as an extension of the builder's own team. Trades represent resources worth developing and appreciating.

Partnering stresses open and frequent two-way communications between builder and trade. Each must share his or her understanding of success and a mutual professionalism. Regular communication can ensure similar expectations. Each should know the needs of the other. Regular communications will alsp help overcome the problems associated with schedule, payment terms, and performance. Larger volume builders will need to use more formal documentation and tracking mechanisms. Smaller volume builders can still sometimes get away with "handshake" deals with some, but not all, trades. It is important to keep a clear and written record of your expectations and agreements. Don't rely upon memory or subjective interpretation about the scope of work involved, the schedule, or payment terms.

In the upcoming chapters, we will discuss the formal documentation necessary to ensure clear communication. Trade contract agreements, policies and procedure manuals, and quality control checklists are a few of the documents we will examine. These documents help define and measure successful performance for all trades.

In today's demanding marketplace, customers look for more than quality construction from their contractors. Today's customer looks for professionalism in design, service, and communication. This professionalism should be exemplified by both builder and trade contractor. In the eyes of the consuming public, there is no differentiation. Everyone working with and for the builder on the job site represents the builder's firm.

Example of a Partnering Principle: "Single Sourcing"
Single sourcing, a term made popular by the Total Quality Management movement, involves using only one trade to do all, or as much of, your work as possible within that trade. Single sourcing basically refers to obtaining labor or materials from a sole (single) supplier (source). Single sourcing is an example of trying to work with and develop a trade contractor.

Single sourcing de-emphasizes the bidding process. It also eases the communication about scheduling as builders can more easily identify which trade or supplier they will be working with on each project.

Single sourcing also allows the builder and trade to develop a more personalized relationship. As they get to know one another, communication and understanding usually improve. Assuming the trade possesses the technical ability to complete the required scope of work, communication improvement has the best chance of helping the builder and trade contractor to successfully accomplish their mutual goals.

Single sourcing allows ease of training as only one trade and the crew will need to be introduced to the intricacies of the builder's construction techniques. Once trained to the system, the trades know what to expect, can work faster, and hopefully make a profit. The efficiencies associated with this level of understanding profit both general and trade contractors.

Single sourcing has its detractors who claim that without bidding, you will quickly be "paying too much" or missing opportunities for improvement by not working with new contractors. It's true that you must be cognizant of the drawbacks associated with single sourcing.

However, I have successfully employed some simple solutions that minimize the downside of single sourcing. For all major single source trade contractors (i.e., HVAC, framing, trim, electric, and custom cabinets), I prepare a quarterly report listing two competitive bids from local contractors for a standard house and the appropriate scope of work or work unit. A work unit could be the square foot framing cost or the per 1000 labor cost of laying brick. By getting quarterly competitive prices from two sources, one can be reasonably assured that the rate being paid is the market rate or better. The same concept can be used to test the prices of our lumberyard by getting quarterly quotes from two additional yards on a simple but complete test home. Using this same home repeatedly as the basis for costing provides a benchmark to measure the rate of price increase/decrease over time. Just remember to always gather the current price for the same set of building components to compare apples to apples.

You can also protect against the loss of opportunity to improve technique or material selections by staying current on construction industry trends and passing this information along to your trades. Better yet, take the opportunity to assist your trades in attending conventions and conferences where such improvements are displayed and discussed. At least, bring suggestions back to your trades as you become aware of them. Don't let them say "we've been doing it this way for ten or twenty or thirty years." That's what the old corner grocery store used to say. Remember them? If not, that's my point exactly.

An Example of Successful Single Sourcing

In Ohio, we built 600 homes a year for many years running using but one HVAC contractor to do all of our work across the state. The HVAC company grew as our organization grew and had 17 trucks on the road daily, many of them exclusively dedicated to our work. It was a positive relationship. They grew to be a successful and profitable firm and we were ensured a trade partner who understood our work, eliminating surprises, cost overruns, and, most of the time, scheduling problems.

Both firms felt they wouldn't want to do without the other. They were our most dependable contractor year in and year out, followed closely by several of our other single source labor and material providers.

The goal of partnering is to transform the trade and general contractor into a true team. Teamwork is essential for success. When the team performs and communicates efficiently, quality, time, and service improve for everyone, reducing budget overruns and increasing profitability.

Good service and high quality result in increased business via referral sales from satisfied customers. Referral sales lower advertising and sales expenses and again increase profits and the ability to competitively price your products. Consequently, the volume of available work increases and the gross profit ratios improve. In general, the builder and the trade contractor both enjoy the benefits of an improved working environment.

Together Everyone Achieves More (TEAM)

What constitutes a team? A team is a group of people who have at least the following:

1. A desire to work together

2. A game plan for success

3. Open communication about the game plan

4. Mutual goals—a common source of motivation

Together is the key word in the TEAM concept. The builder/trade contractor relationship works much like a football team. The coach establishes the game plan. The quarterback orchestrates the field offense and depends upon the lineman for protection; the lineman depend upon the backs and ends to catch the ball. If the lineman and backs fail to work together the game plan disintegrates. A football team can have the best offense in the nation and still fail to win without an equally effective defensive team.

In our industry, the builders, developers, and their organizational management are the football coaching staff. The construction manager or field supervisor is similar to the quarterback. The trades and suppliers correspond to the position players. The customers are the fans. If your team works together (doing all the things outlined above), it will perform according to the game plan and will please the fans. The fans will consider their money well spent. The fans—your customers—will speak highly of your performance inspiring other people to come to the "game" next week.

If, on the other hand, your team's effort is full of errors, lacking both communication and a consistent plan to achieve the common goal, the lack of coordination will be evident to all. Seeing your players continually "fumble the ball" or lose to your competitors, customers will not come back. Worse yet, they may come back with their lawyers.

The Concept of Comfort Zones

Every person, as an individual, has an area within which they feel comfortable and operate most efficiently. Most people are comfortable with family and friends and begin to stretch their comfort zones when they begin to interface with strangers. Figure 1.1 depicts a person surrounded by an area where he or she most often and most comfortably operates. Outside this comfort zone are the areas he or she crosses over into less frequently and with less comfort.

Some of the less comfortable situations you may personally identify with may include the following:

1. First dates

2. First meeting with girlfriend/boyfriend's parents

3. Job interviews

4. Public speaking opportunities

5. Negotiating with a new trade contractor

6. Dealing with dissatisfied customers

A tremendous degree of elasticity can be found within the comfort zone. It can grow as the individual learns to be comfortable in new areas. In the above examples, each provides the opportunity to become more comfortable as one gets accustomed to new in-laws or a new job. One learns that with practice the challenge of public speaking can become more tolerable or that a new trade will hopefully become a proven business "partner."

As one's comfort zone grows to include a new situation so does the individual's efficiency of operating within that particular area. When you are within your comfort zone, stress and fear of the unknown do not distract you. You can concentrate on planning and performance, on enjoying and understanding. You must be comfortable to have the clear head and understanding that allows for planning and visualization.

As individuals have personal comfort zones, organizations also have professional comfort zones. By understanding the concept of comfort zones the builder/trade contractor/supplier team can grow in several ways. Experience and professional development can expand their comfort zones. Education, training, and seeking opportunities for new challenges will

FIGURE 1.1 Comfort Zone Concept

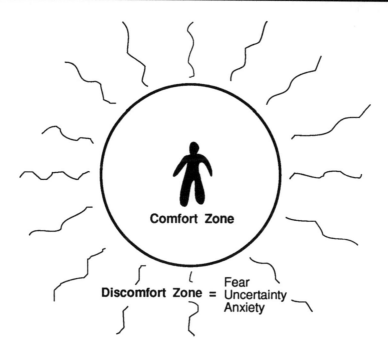

Comfort Zone

Discomfort Zone = Fear
Uncertainty
Anxiety

stretch builders and trades past their current comfort zones and eventually allow their comfort zones to grow. Professional development is a key to long-term success for any building organization as employees and trade contractors learn and progress.

The immediate rewards of understanding comfort zones include maintaining quality standards, improved construction times, and increased profitability. By understanding and keeping trades and suppliers working within their comfort zones, the builder will be contributing to everyone's success. How? Because within their comfort zones is where the trades will do their best and fastest work. Within their comfort zones, problems are less likely to exist or be perceived. Confidence and teamwork flourish, and the contractor has the best chance of meeting and exceeding goals.

In Chapter 3, we will look at some of the details of the general comfort zone parameters within which the different trades, including suppliers, operate best. This discussion includes methods for encouraging trades to stretch and expand their personal and organizational comfort zones. See also Appendix B for sample comfort zone checklists.

Who's the Boss?

Many of you may be thinking that, so far, this sounds like a bunch of "pie in the sky" rhetoric and not like what goes on in residential construction in my neighborhood. All this discussion of partnering is too goodie-goodie for what is really a cutthroat business. In fact, trade contractors of today are different from their predecessors. The culture of business in general is different. Our industry is usually slow to adapt to new business trends. This is certainly true in the case of contracted labor relations.

Most of the U.S. business world has been using the techniques discussed in partnering for years now. It was at the heart of the quality improvements noted by the Ford Motor Company in the 1980s. It is at the heart of the decline in power and influence of the labor unions as a whole. As business becomes more humanistic in approach, customers most often reap the benefits. After all, people are our customers. When we become more people conscious, a side benefit is customer consciousness.

You are right, however, ours is a tough industry. Generally, trade contractors are independent in nature. Why else would they risk everything to work in a business that is hard work, dirty, in an uncontrolled environment and subject to the whims of the weather? Labor shortages have caused most trade contractors to hire helpers who are undertrained and unprofessional in demeanor. Apprenticeship programs don't exist in the volume that they used to.

In the chapters to come, I will give examples of when the tough approach has been the right solution for my firm. I'll also point out that more often than not, the concept of partnering has been the long-term solution to most labor-related problems I've become embroiled in.

The builder or general contractor does not lose power or control of the construction by using partnering. Conversely, when properly applied, the partnering techniques enable the builder to maintain greater control through a very positive relationship with his trades. Trades will fight to keep your business, even in a busy and competitive marketplace.

Many builders take the attitude that "this trade will perform or I will find another for my next job." Sometimes this attitude is unavoidable, but often it is an excuse for the builder not doing a good job of recruiting, selecting, training, and communicating properly with the existing trade contractor.

Many builders assume that they have complete control over their jobs. In reality, builders are often at the mercy of their contracted labor, and whether the builders admit it or not, the trades often are in the bosses role within the relationship.

The teamwork approach de-emphasizes the boss role and emphasizes the leadership skills of the builder. The role of the boss has changed in many ways over the past two decades. Today, a successful boss is a leader, a teacher, and a coach more than a discipline-oriented dictator. As a builder or general contractor, are you ready to assume this role? Are you ready to treat trade contractors as part of your team?

As an Example: Who's in Control?

For over a year, my schedule has undergone numerous changes as a result of our roofing contractor not being able to perform as scheduled. It had deteriorated into us having to give him notice 30 days before we needed him to ensure a mutually compatible schedule.

This required us to give scheduling notice oftentimes before we began construction. It required two additional telephone calls than what we thought necessary. The tight labor market and the fact that the workmanship was excellent from this contractor caused us to put up with the inconvenience for many months.

In essence, this put the contractor in control of my schedule. I, normally, had to slow down my schedule to accommodate his desire to stay 30–45 days out. Numerous discussions of the pros and cons of their system resulted in no change.

Finally, we informed the roofer that we were going to have to go elsewhere for roofing labor because of the schedule impact. As this is being written, we are negotiating some method ensuring our place in his schedule on a 20-day notice schedule. (Twenty days being approximately the average time from construction start date until roofing labor commencement.) Unfortunately, we are also negotiating with a second source of roofing labor if we can't reach a mutually satisfactory agreement with roofer number one.

2

Finding Trades Who Can Do the Job

The most common complaint among residential builders in good markets is that there are never enough quality well-trained trades. One reason for this is that builders often fail to plan ahead. They look for trades only when they need them. This just happens to be the same time the competition needs them. With so many builders trying to put them to work, the trades are just too busy to be looking for new sources of work. Shocking concept, isn't it? And builders wonder why some of the trades seem so independent and indifferent.

How to Find Trades and Still Be Home by Midnight

Make efforts to expand your trade contractor base when you *don't* need them. Interview and negotiate when business is slower, preferably in the winter months for those who build in the colder climates, and during the slower parts of the year elsewhere. The trades are less busy, their competition is greater, and they are more willing to meet and negotiate. Searching ahead gives both the builder and the trade an opportunity to communicate and plan strategy for the coming building season. Recruit trades before you need them!

The next task is to discover *where* to find trades. The following simple methods are worth trying:

- Visit **other builder's** subdivisions and job sites and see who is working for them. This is an especially good method of determining quality and schedule duration standards.

- Ask **material suppliers** whom they would recommend in their specific area of expertise. Ask the brickyard for recommendations for masons, the roofing supply house for roofing contractor names, etc.

- Advertise in the local **newspapers.** Utilize a trade contractor application form similar to an employee job application form. (see example in Figure 2.1.)

- Search the associate member directory of your **local home builder's association.** Membership in the association indicates, at least, a minimum appreciation for professionalism and teamwork within the industry.

- Ask **current trade contractors** for contractor references in other trades. They will generally provide good recommendations on trade quality. They will be especially good references for trades that they follow in sequence of construction. The recommending trade will be protective of his comfort zone.

- **Banks and other lenders** will often have recommendations on trade contractors via their construction inspectors.

- Check with **local trade contractor associations.**

- **Building inspectors** from local government jurisdictions are also good sources of information on quality trade contractors.

- Check the **yellow pages.**

If these simple recruitment methods are not working, other "drastic" steps can be attempted to increase the available trade base in your geographic area. The following methods generally are advantageous to larger volume builders or to a group of smaller builders working together through their local home builder's association:

- Advertise for trades in the major metropolitan newspapers of areas of the country where the markets are soft. Many builders from the

FIGURE 2.1 Trade Contractor Application Form

Date _____

Contractor Name: _____

Address: _____

City: _____ State: _____

Zip: _____ Phone #: Office/Home _____

Fax: _____ Mobile: _____

Local Contact: _____

☐ Trade Contractor ☐ Supplier ☐ Both

Type of Organization:

☐ Corporation

☐ Partnership   Federal Tax I.D. # or SS# _____

☐ Proprietorship   SS# _____

Principles or Officers: _____

_____

_____

Liability Insurance Company: _____

   We are required to keep on file a "Certificate of Insurance."
   Received: ☐ Yes Liability Limits _____

Do you have any employees (full-time or part-time)? ☐ Yes ☐ No

If yes, do you carry Workman's Compensation?

☐ Yes ☐ No   Policy #: _____

(Please provide a copy of your Workman's Compensation Certificate and a renewal certificate each time it is renewed.)

References:

(1) _____ (2) _____
    NAME: NAME

_____   _____
ADDRESS: ADDRESS:

_____   _____
CITY/STATE/ZIP: CITY/STATE/ZIP:

Bank Reference: _____

Trades you are currently performing: _____

_____

Recruitment calls can tell you volumes or "Here today gone tomorrow"

As you start calling around to recruit the potential trades, take note of how easy or difficult they are to reach. Do they return calls quickly? Do they have both pager and cellular
phone? Do both methods get a quick response?

If the trade isn't easy to reach now, they will not be easy to reach when they are working for you. (A note on pagers: I have had some experience with trades who respond
immediately to a telephone number that they do not recognize on their pager. They
think it may be a new business opportunity. That same trade may ignore your page when
they recognize the number as being from a builder they are working with and they don't
have the answer you want to hear.)

All too often, we overlook the difficulty encountered in initially trying to reach a potential trade only to hire that individual and find that they are increasingly difficult to
communicate with.

Rocky Mountain states and their associations recruited trade contractors from California where the market was soft in the first half of
the 1990s. Texas, Oklahoma, and the other energy industry states
were experiencing a sluggish market during the last half of the 1980s.
This was when the coastal markets were booming and the energy
state's markets were dead. Many 1986 job sites in Atlanta, and
throughout the southeast, were surrounded by pickup trucks with
Texas and Oklahoma license plates.

■ This concept can be expanded to include recruitment from markets
  not necessarily dead, but where the wage and subcontract labor price
  are notoriously lower than in your area. By offering a work volume
  higher than currently received wages, the trade contractor can often
  be convinced to relocate. This strategy requires great care to guard
  against only the marginally competent trade contractors responding
  to the advertisements.

■ Finally, you can try what I call the "equity assisted" trade contractor
  recruitment strategy. Using this strategy, the builder literally goes into
  partnership with a potential trade who needs some capital and perhaps some planning and business organizational assistance to get
  started on his own. These potential trades are often crew members
  of an existing trade and great care must be taken to prevent ruffled
  feathers. The advantages of this method are that the builder gains a

degree of inherent loyalty from the trade contractor and that training can often be personalized for the builder's products and construction methods. Remember that to qualify as a true "independent contractor" in the eyes of the IRS, several qualifying tests must be passed (see Chapter 1). For example, the IRS requires that the contractor cannot work solely or exclusively for one general contractor.

Finally, when you look for trade contractors, the absolute worst place to advertise and interview is on your existing job sites. To advertise "carpenters needed," from a piece of 4 x 8 plywood on a job site is a terrible mistake. Production at that job site will be slowed by every passing, unemployed Tom, Dick, and Harry who has ever hammered a nail. The builder or the construction supervisor will be interrupted continuously. The advertising builder is announcing his lack of professionalism and admitting weak pre-planning. Furthermore, any trade contractor worth hiring should be working or looking for work in a more professional manner.

Production builders, who have a production office setup on site (i.e., a trailer), may be able to do some on-site recruitment effectively. This, however, must be monitored and evaluated perpetually to ensure that project productivity isn't compromised by frequent interruptions by unqualified applicants.

Evaluating Potential Trade Contractors

Once you have recruited several qualified trade contractor candidates, the real challenge begins. Selecting the "best" trade for a project, a series of houses, or as a single source supplier requires an in-depth review of three sets of criteria. The selection criteria include business competency, trade competency, and price.

These evaluation criteria parallel the old business definition that every product producing entity in the world offers three basic ingredients to the consumer: price, quality, and service. The old saying used to be that of those three, the customer could pick two. Today, consumers are better educated and more product savvy. Competition has encouraged product improvement and a greater appreciation of customer service. Builders in the next century will be pressed by their customers to offer all three ingredients. This will only be possible if trade contractors offer the same to builders.

The most effective way to evaluate a potential trade contractor is to interview the owner of the firm. Visit the trade's office if applicable. Note the attitude and philosophy toward management and labor. Try to question the goals and ambitions of the firm and its owner. Match those against your goals and ambitions. Compatibility is critical for teamwork and for the establishment of mutually acceptable expectation levels. Keep the three basic sets of criteria in mind when interviewing and selecting trade contractors and suppliers.

Business Competency

Business competency is an increasingly important ingredient in the light construction industry. Note the attitudes of employees and telephone contacts within the trade contractor's operation. Constantly look for the indications of professionalism. These can include:

- How long has the trade been in business?

- Does the trade maintain an office? Does he use an answering machine, mobile telephones, radios, or other communication technology? Does he employ a receptionist, dispatcher, or secretary with whom you can communicate during normal business hours? Does the trade have printed business cards, quote sheets, or business stationery? Will the trade provide you with an invoice for service? Can the trade understand your purchasing system if you use purchase orders?

- Does the trade have evidence of liability insurance and worker's compensation insurance coverage? Limits of coverage can be an indication of the size and sales volume of the organization. In today's market, trades must have their own coverage, and builders are strongly recommended to require and maintain current certificates of liability and worker's compensation insurance.

- How are the trade's credit references? Check suppliers of materials, several if possible, for the status of the trade's account(s). Credit bureau reports can also be a source of financial information. Can the trade secure a bond? If so, the amount is often an indication of financial credibility and operational maturity. The key is to determine if the trade contractor can and will pay his bills on time. The last thing a builder needs is the headache of an overextended trade causing liens to be filed upon the builder's jobs.

- Is there evidence that the trade contractor and his employees understand that personal appearance has a significant impact upon professional image? Remember that everyone on a builder's job site represents the builder and his firm. In the eyes of the consuming public, that trade contractor is your employee.

These general business criteria characteristics must be reasonably well covered to determine whether the trade contractor will be easy to work with

The Case of the Missing Insurance Certificates

A consulting client of mine had a policy to require certificates of liability insurance and worker's compensation insurance before hiring a new trade contractor. They did an excellent job of consistently applying this requirement to each contractor hired. They were happy to have such a good system in place; their insurance company premiums were lower than before.

Unfortunately, they did not require each trade to give them an updated certificate each year or 6 months depending upon the term of the policy. Over the course of several years, all of the certificates on file had expired. At just the wrong time, their insurance company's auditor showed up to perform his audit and to calculate the additional premium. (Don't ask how to calculate the additional premium, it's almost as complex as calculating the cost of a building permit in some municipalities.)

When my client tried to get updated certificates from his suppliers and trades, he found that about 50 percent of his trade contractors had let their policies expire. The audit premium ended up costing the builder over $5,000 additional dollars.

Make sure that your system allows you to track policy expiration dates. Set up a form letter to send out 30 days prior to policy expiration requesting a new certificate from the trade. After the expiration dates, don't release checks to the trade until the certificates are received.

and will create no new problems. Ensuring ease of communication for scheduling and service needs and determining compatible business goals and philosophies are the primary objectives in evaluating business competency.

Trade Competency

Trade competency refers to the contractor's ability to deliver a quality finished product in a timely fashion. To verify a new trade contractor's quality, inspect past work. Looking at previous projects works well for the finishing trades but much of the work of the rough trades is hidden in finished products. To inspect the quality of work for the rough trades takes a commitment of time and effort, but generally, it is worth the exercise.

Remember that the quality you seek depends upon the type and, often, the price of the product. Make sure you are comparing apples to apples. For example, if a trade usually works on $100,000 houses and you build $500,000 houses, you should closely compare the trade's quality with your expectations. The quality of some trades (especially trim carpentry) depends on a clear distinction between pure production and true custom construction.

Several alternatives and subsidiary actions to personally checking the quality of the potential trade contractor exist. Builders can inquire about the workmanship from the following sources:

- Other builders, especially HBA members

- Inspectors, both government and bank

- Other trade contractors, especially related trades

- Homeowners

Homeowners? Yes. Interviewing owners of homes in a subdivision where a potential trade contractor has done the electrical, plumbing, HVAC, insulation, etc. can give a builder an indication of what his customers will be thinking of his homes, several years from now, if you use the same trade. Of course, design, material, specifications, and construction supervision also influence a home's quality. However, through talking with owners, a builder can learn much about a trade's reputation for customer service, timely construction, and professional attitude.

Beyond the quality of work, several other trade characteristics should be addressed when recruiting new trade contractors. These include the following:

- Trade license—Do the trade contractors have the appropriate trade license? Licensing requirements are localized. Business licenses are different from trade licenses. A licensed or certified contractor normally has passed a test or completed an educational program designed to document a minimum set of trade competencies. Local building departments can tell you which trades need which licenses beyond a business license in their jurisdiction. These licenses are required by law in many states.

- Timeliness of construction—Have the trade contractors, from previous references, been able to meet the scheduled time deadlines placed before them? Were they on time? Did they finish 100% complete on schedule? Was the duration of his scope of work within standards? How much lead time is necessary to ensure having a crew available? Did the trade perform with a full crew, finishing the job before moving on to his next? Did the trade work weekends when necessary? Did the trade work over eight hours per day when necessary? What is the trade's policy on holidays? What is the trade's policy on weather related conditions? All of these questions will give you an idea of schedule compatibility.

- Timeliness of warranty service request work—Does this trade contractor understand the importance of prompt service response to customers under warranty?

- People skills—Does the trade contractor understand the builder's and customers' need for a professional level of communication and record keeping? Do service technicians display tact and understand the importance of loyalty to the builder and the product?

Quality of workmanship, timeliness of work and service, and professional approach to customers, construction managers, suppliers, and inspectors are important trade ingredients to be measured. The answers to questions about these ingredients will give you a reasonable view of the trade contractor's technical competency.

Price

Many builders consider price the single most important criteria in selecting a trade contractor. However, potential hidden costs can easily cause an initial low bid to increase. Rather than consider price as *the* most important factor, look for a balance between business, trade competencies, and price.

A low price is not always the best value. When you analyze the potential costs of service on mediocre quality, schedule delays from understaffed or poorly managed trades, and customer complaints from a nonprofessional approach to communications and business procedures, the bid price has implications far beyond the quote. It is hard to place a value on customer satisfaction or the lack of it. Every satisfied customer tells five potential customers of his satisfactory experience. Every dissatisfied customer tells ten times that number of his negative experience. How many builders can survive that type of negative publicity? The marketing cost savings of one satisfied customer referral can amount to 3 to 5 percent per referral sale.

In summary, the trade contractor recruiting criteria should match the three things that every builder strives to offer his customers: a competitive price, quality construction, and prompt and courteous service. The builder, as the customer of the trade contractor, should recruit and evaluate potential trades with these elements in mind.

Negotiations

Eventually, almost every trade contractor recruitment involves some degree of negotiation. Robert Maddux, in his book *Successful Negotiation*

(Crisp Publications, Los Altos, CA, 1986), defines the process this way: "Negotiation is the process used to satisfy our needs when someone else controls what we want." Negotiations are necessary anytime an aspect of a potential transaction becomes unsatisfactory to the builder. Negotiations with trades can include, but are not limited to, price, scope of work, duration and schedule of construction or delivery, payment terms, warranty, issues of job site preparation and cleanliness, and communications with customers.

Applying the partnering philosophy, presented in Chapter 1, to the negotiation process requires considerable effort and preplanning. To prepare for a successful negotiation, a builder should do at least the following:

- Have a clear understanding of what his objectives are in the negotiation process.

- Get to know the trade contractor and his background.

- Discuss with the trade his goals and operational philosophies.

- Find out what motivates the trade contractor. What are his hot buttons? Some trades want and need to work six days a week and routinely work holidays and overtime. Other trades may have a more laid back approach and can be motivated by a builder who understands the need for deer hunting season schedule extensions, etc. Some trades need more frequent and rapid payment terms to meet payrolls and overhead expenses and builders can negotiate better pricing in return for quicker payments. The list can go on and on.

- Prepare a list of the issues needing definition and be prepared to negotiate if necessary each of these points. A partial list of sub-issues frequently requiring negotiation would include: size of crew, schedule lead times, liquidated damage clauses on commercial or large scale residential projects, disposal of scrap or trash, music or noise, etc. Don't forget to plan for the small, seemingly less significant items.

- Define the areas of probable disagreement and be prepared to spend additional time in resolving these areas. Special preparation is often necessary in these conflict areas to be certain of reasonable compromise solutions. Every conflict area is worthy of two or three potential solutions. Planning is the key to finding the solution that will meet both parties (remember, win/win) objectives without resentment or lasting conflict.

- Review potential solutions for areas of disagreement by remembering that compromise requires give and take. Plan to "give" on points where you will be getting or taking something of equal or greater value (to you).

Each point of negotiation should be reduced to writing and placed in the subcontract agreement. This process of preparation and negotiation serves the additional purpose of allowing the builder and the trade contractor to set mutual expectation levels on all negotiable issues. Guessing and "shooting from the hip" are greatly reduced. A professional atmosphere exists and prevents many of the time-consuming problems that often surface during construction.

Trade Contractor Agreements

Every builder should utilize written trade contractor agreements with all trades (see Figure 2.2 for an example). Without written agreements builders leave themselves exposed to a myriad of problems, including defining scope of work and quality performance, debates over schedule duration and lead times, payment terms and timing, warranty work, customer service, and a host of other potential problems. Written agreements allow for some legal protection but the big advantage is communication. Like a construction schedule, trade contractor agreements offer the builder and the trade the opportunity to analyze potential problems on paper before they happen. This allows mutual expectation levels to be established in a short period. Working with trades on handshakes lacks professionalism and causes problems because of differences in expectation levels.

Trade contractor agreements can be prepared for each house being built. However, this is efficient only for builders of a few very large or highly customized homes. Builders who build from a set of standard models are better off having trade agreements for a specified period of time or number of houses.

Trade contractor agreements usually include the following three sections:

General Section

The general section will probably be the largest section of the agreement. It should include information common to all trade contractors such as information on insurance, bonding, and licensing requirements (where necessary). It should also include boilerplate wording on a trade's knowledge of building codes, plans, and specifications. The general section should empha-

FIGURE 2.2 Sample Contract Cover

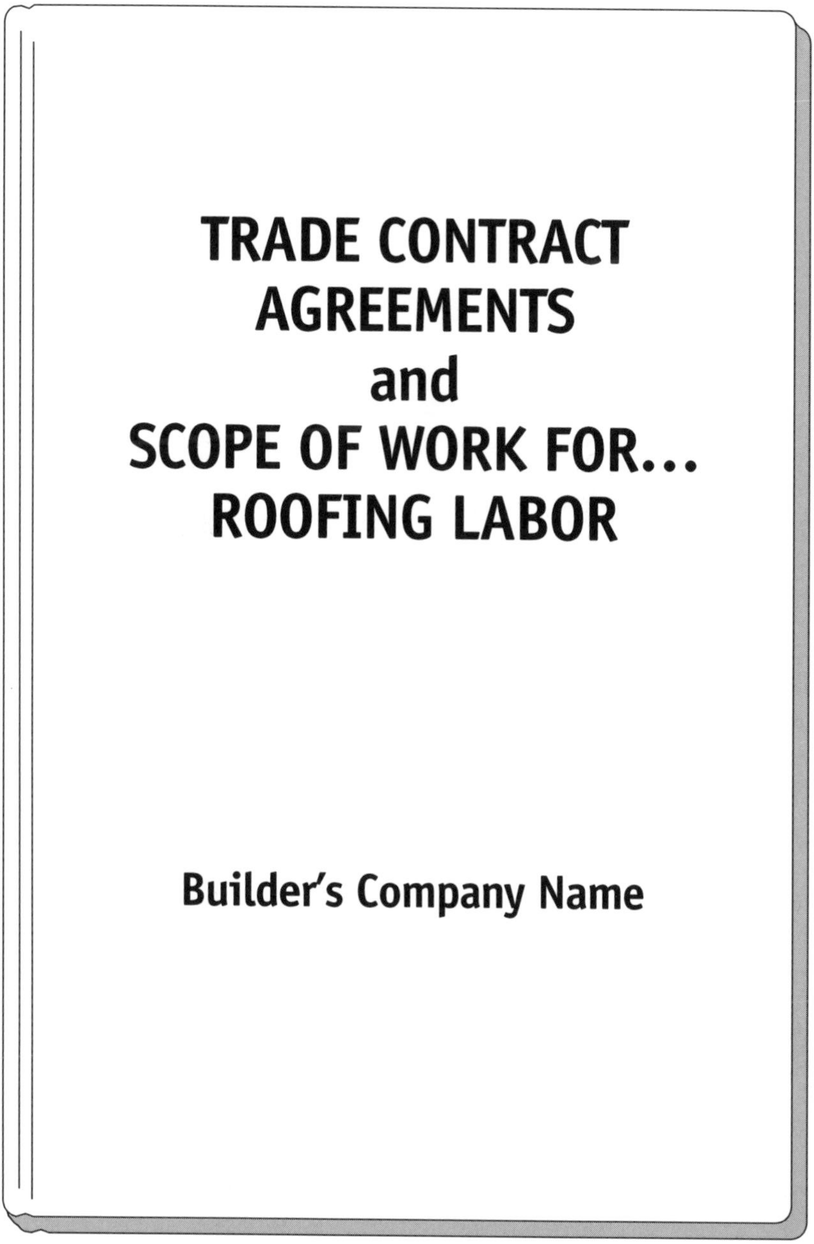

TRADE CONTRACT AGREEMENTS
and
SCOPE OF WORK FOR...
ROOFING LABOR

Builder's Company Name

size responsibility of general site conditions such as damage, acceptance of work by preceding trade contractors, theft, use of drugs and alcohol, and site cleanliness and trash disposal. Also to be included are statements on compliance with building inspectors, OSHA regulations, and even management quality control inspection procedures. This section should also state the trade contractor warranty.

Trade-Specific Section or Scope of Work

This section addresses the scope of work for the trade in question. It should detail everything included in the bid price. Use this section to add labor detail to the referenced construction and material specifications. Detail materials or supplies that trade contractors are to provide. Where there are several common industry methods of construction identify the desired method. For example, if a builder prefers that all drywall be screwed as opposed to nailed, this is the section to detail that expectation. Who hauls off the dry-

---

Could I get a draw today?

I had a brick mason named Willy. Willy and the poor boys is what we called him and his crew. Each week they had more excuses than a kid without his homework about why he needed to get paid today for work he was performing tomorrow. Sound familiar?

Without a good set of documented procedures, it becomes personal when you can't or don't want to pay ahead of schedule. The larger the building company, the more important the need for disciplined and documented procedures. Make it a hard and fast rule. Payment for work completed and approved by a certain date or day of the week occurs the next regular payday. I have had the best results paying labor contractors weekly, on Friday, for work completed and approved by Wednesday at close of business. Payment for any work completed after Wednesday goes out the next week. A larger builder may need to cut off on Tuesday at close of business to ensure enough time to process the checks.

You must be firm and fair on this policy. It must be clearly stated and understood before the first job begins. For trades with scope of work durations exceeding 2 to 3 days, it may be necessary to break the work payment into several draws or purchase orders.

Avoid the mistakes that I have made by being soft on the policy. Willie and the poor boys got paid the final draw on a job with about a day and a half worth of brick work left. The next time I saw them was two weeks later and they were looking for more work (I had to bring a second mason in to finish the original job). They had lost all their money gambling and were desperate. Unfortunately, the irresponsible few make it necessary to have firm policies for all.

wall scraps? Who does the roof flashing and the felt paper? Does the builder pay for a crane to set the trusses? These and similar questions need to be addressed in these trade specific sections. There should be a different trade section for every trade contractor used.

Payment Terms Agreement Section

This part of the agreement is the all-important price and payment section. Payment schedule and discounts should be detailed, as should the procedures for having work approved for payment. If the builder utilizes purchase orders, detail the procedures in this section. If variance purchase orders are utilized, these generally require special mention.

The terms section should state the duration of the agreement, which can be a period of time or a number of houses. The most appropriate method will be determined by careful review by builder and trade contractor together. Remember the win/win attitude.

Appendix A presents an example of a complete trade contractor agreement. You should not use this example without customizing it to your situation. The sample is provided as a thought-provoking foundation from which a contract specific to your operation can be developed. Remember to have an attorney check your draft contract to ensure conformity with state and local contract law.

Remember that the legal issues surrounding contracts are sensitive to many state and local conditions and requirements. Builders should never just use a boilerplate contract without 1) careful review and editing by legal counsel who is construction-knowledgeable and 2) specific tailoring to fit the operational philosophies of the company. Additionally, the builder and existing trade contractor(s) should review the proposed agreement and look for mutually agreeable additions before finalization.

3

The Trade Contractor's Comfort Zones

The first chapter introduced the importance of keeping trade contractors in their respective comfort zones. This idea is based upon the concept that everyone works best in an atmosphere of comfort and understanding. How do builders ensure that their trade contractors stay in their comfort zone? The question has as many answers as there are trades on the builder's job sites. Many trades have needs that other trades do not. By taking each trade's needs into consideration you can help keep them in their personal comfort zone. This section lists a few of the areas that trades from the major areas of residential construction have said keep them working to their peak efficiency. Providing these things and the resultant atmosphere is the key to motivating trade contractors. See Appendix B for sample Comfort Zone checklists.

Minimum Requirements

Common to All Trades

- A detailed scope of work negotiated prior to commencement of work.

- A standard payment terms agreement, often tailored to the trade's special needs.

- A clean job site with adequate access. This includes an area to park and mud-free access to the site and to their materials. In the wet times of year, increase your gravel budget to allow for this environment. There should be a driveway to the site and a turn-around extension wherever possible.

- Properly scheduled lead times (limiting emergencies). Set up a scheduling system that red flags the work order date at the appropriate interval prior to labor being needed on site.

- A detailed set of plans.

- Builder adherence to the payment schedule. My favorite labor payment schedule is scope of work agreement stating that work completed by Wednesday at close of business gets paid that Friday.

- Regular and specific communications on performance, schedule, and especially change orders.

- A construction manager who understands when the job site is actually ready for the trade to begin. Material drop sites must be marked and planned to be where the trade needs the materials. Over-digs on basements must give the trade room to do his job. All rough door openings must be accurate before your trim carpenter can install doors and be in his comfort zone.

- A good method of communication to the construction manager or builder.

- A porta-john available until the plumbing is operable within a home. This saves you time and money by keeping the trades closer to the job site and keeps your home and customers happier by eliminating the temptation for your trades to foul the site.

- A dumpster or refuse containment system of some sort on site throughout the construction process. A trash can inside the home, once the walls are framed can also help with job site cleanliness.

Excavators

As the first trade on most sites, the excavators need close supervision as the excavation begins. Make sure that the excavator double checks the lot dimensions, set-back requirements, and proposed drainage systems. To do their job effectively, excavators need the following:

- Building layout staked to actual size.

- Over-dig measurement 24" minimum or as agreed.

- A detailed excavation plan with grade depths, step-downs, etc.

- Superintendent or construction manager on site when excavation begins to ensure that communication is clear and to answer questions.

- All stairwells and chimneys staked out.

- Natural drains or sewer drains planned and drawn out.

- Utility trenches clearly marked and plans detailed.

- Proper lead time to schedule underground utility protection inspections and markings.

- All lot pins clearly marked.

- Good coordination with management on backfill timing and requirements.

Footer Crew: Basement Foundation

- A dry hole or work area.

- Access to hole via open corners and a walk path around excavation site.

- A level surface and virgin soil.

- A detailed drawing of all step-ups, bulkheads, and level changes.

- Rebar and foundation tie material available on site.

- Easy concrete truck access to the footings.

Foundation Crew: Block or Poured Basement or Crawl Space

The foundation crew, if different from the footer crew, should pay close attention to the footer crew's work. Foundation walls that are an out-of-square footer will often have to be taken down. The things a foundation wall crew requires are as follows:

- Square and level footings of proper width.

- A dry hole with a sump pump available.

- Enough of an over-dig to ensure room to work.

- Temporary electrical power for sump.

- Location of block and mortar. Block at various points and proper quantities in the basement, garage, and at grade.

- For poured walls, easy access around hole for mixer truck and booms prior to placement.

- A detailed drawing of anchor strap or bolt placement.

Foundation Crew: Concrete Slab

- Lot pins or survey pins clearly marked.

- An adequate supply of block, if block and fill slab.

- An adequate supply of form material if monolithic slab.

- Pad built to proper size and soil compacted to needed strength.

- Plumber coordinated to follow closely behind footer and form crew.

- Any under slab electric coordinated to not delay slab pour.

- Concrete either ordered directly by concrete flat work trade or on will call with ready mix company.

- Anchor bolts and insulation available on site.

- Anchor bolt placement diagram.

Rough Carpenters or Framers

Framers are one of the most vital contractors for most builders. There is a continuing shortage of skilled carpenters in most markets today. You want to be the builder they want to work for. Extra care must be taken to maximize the carpenters chances of success on your jobs by having the following in place:

- Electrical power closely accessible to the house.

- Easy access to lumber drops.

- Lumber drops in stages of construction (i.e., first floor, second floor, trusses and sheathing, windows and doors).

- A last on/first off stacking pattern of materials from the lumber yard.

- Square and plumb foundation walls or slab.

- On basement foundations, carpenters prefer to work where the backfill has been done and the backfill completely filled for ease of moving around site. Of course, excavators prefer to do the backfill after the framing is done to add support to the foundation walls during the backfill sequence. The builder's "team" must determine the best methods for their operation.

- Material lists, cost sheets, and layout diagrams (detailed with intended usage) should be available. This is a key point to both trade contractor efficiency and job profitability. A detailed list of how the 2" x 10" x 14' material (perhaps floor joists) is to be used and the number of the component needed in that dimension should be clearly stated.

- Brick ties should be available for the framers if brick is to be installed.

- Roofing felt and nails should be available (if this is in your framer's scope of work).

- A realistic nail and adhesive allowance should be built into the contract or you should provide an adequate supply of nails and adhesive in the correct sizes.

- You should conduct a meeting with the superintendent prior to the framing.

Heating, Ventilation, and Air Conditioning Contractors

- Proper lead time to study plans and order size and style of unit(s) needed for job.

- Clearly marked set of plans showing heated and cooled area and special conditions.

- Ability to get onto job ahead of plumber and electrician to ensure efficient heat and air return runs.

- Dry basements with concrete poured is ideal (crawl space of at least 32" from ground to floor system in crawl space environments).

- Temporary electrical power in place.

- Gas line clearly marked.

- Clean work areas.

- Some HVAC contractors also want the house to themselves. Otherwise, plumbers and HVAC contractors can often work together.

Plumbing Contractors

- Plumbing walls and chases properly framed (2 x 6 walls where needed, etc.)

- Plumbing fixtures available at right time per predetermined schedule.

- Water heater location clearly marked on plans.

- Floor drains in place.

- Exterior hose bib locations marked on prints.

- Under slab plumbing locations clearly marked on foundation prints.

- Builder should double check sub-slab plumbing locations prior to pouring concrete.

- Have sewer stub and gas line (where applicable) located prior to initial plumbing visit to site.

Electrical Contractor

- Mark door swings on subfloor or rough door openings to ensure electrician knows where to place switches.

- Provide detailed electrical schematic to eliminate confusion.

- Allow electrician to meet with homeowner (where applicable) on first day of rough to mark extra outlets and negotiate switch placements, etc.

- Negotiate or communicate who does cable TV and telephone jacks prior to start of each job to avoid confusion. Decide on quantity and location.

- Make sure all wires have been pulled through the drywall. The electrician is not noted for his ability to find buried wired without making large holes in drywall.

- Check electrical box placement and depth prior to insulation. Make corrections before drywall.

Insulation Contractors

Insulation installers are always on a tight schedule. They must work efficiently to make money. At this point, the builder is usually in a hurry to get to the drywall phase. To work with the maximum speed and efficiency insulation contractors need a job site with the following:

- Rough mechanical inspection complete.

- All cavities open for insulation (i.e. spaces between end and rim joists, corner framing, under cantilevered bay windows, etc.).

- All backing installed.

- Windows installed and shimmed.

- Mechanicals completely roughed.

- All framing repaired after mechanical rough.

- Drywall not stocked in house.

- Attic access holes, medicine cabinet openings marked, and curbs framed.

- No other trades on site.

Concrete Flat Work Contractors

Concrete flat work can be done at several points in the construction process. Interior work can be done as part of the slab foundation or in the case of a basement, it can be done prior to framing or after framing as long as basement bracing can be installed and not removed. Exterior work can be done at several noncritical stages during the second half of most residential construction projects. Flatwork contractors need the following to be most efficient:

- Access to basement if framing complete before pouring basement floor.

- Dry work area.

- Crawl space pours, where applicable, scheduled prior to framing.

- A 4" ledge for support of poured garage floors provided by masons. Reduces concrete service calls when interior backfill settles.

- A diagram showing requested control joint placements.

- Sump crock for drain tile in basement and pump available or on will-call for installation as basement floor is prepped.

- An ample supply of form material, especially for unique or curved outside drives and walks.

- Pipes under drive and walks.

- A comprehensive diagram with measurements of outside concrete (driveway and sidewalks).

- Width measurements on the outside concrete. These can vary greatly from project to project.

Brick Mason

All exterior siding contractors work in very difficult situations. They are exposed to the elements, have to climb scaffolding to do much of their work, and are very dependent upon an adequate supply of materials. The following will help the brick masons overcome these adversities:

- Have basement homes back filled prior to starting brick.

- Have good rough grade.

- Have enough lintels of the correct size to head off all windows, doors, and garage doors.

- Stock brick and mortar in several locations around the house where brick is to be used.

- Confirm that all brick is from same lot number.

- Have brick ledges should be installed by foundation or concrete crew during their stage of construction.

- Confirm color of mortar and brick and joint type (concrete, flush, raked, etc.) prior to delivery.

- Have quoins and keystones and vents, etc., clearly marked on prints.

- Have all windows and exit doors in place before mason starts.

- Have eave and decorative vents must be on site for mason to install at proper time.

- Don't apply shingles to porch roofs, if brick will continue above porch, until after brick work is done.

- Let mason know if fireplace chase is to be brick or siding. This is especially important on one-story homes where extra scaffolding may only be needed if chase is bricked.

Drywall Contractors

Good drywall depends on the quality of work performed by the framing contractor. Drywall finishers suffer if the hangers have to piece together a wall section with scrap material. The hangers should know the hanging pattern (horizontal vs. vertical). Board is available in a variety of widths including 54" for 9-foot ceiling heights. Purchasing the correct boards will go a long way toward making the drywall job easier. In addition, you can provide the following:

- Backing nailers installed on all surfaces to be drywalled.

- Detailed board layout on plans (saves time and money by decreasing waste).

- Board stocked in unit properly. A window, with removable sashes, should be available for stocking drywall on the second floor of each two-story home. If you provide the board, make sure the delivery driver has a list of what size board is used where to prevent contractor from having to move a load.

- Framing double check.

- Dumpster conveniently located and not full.

- All house details framed to correct rough opening size (i.e., attic access holes, medicine cabinets, stairways, niches and special ceilings).

- Clean floors and garage area.

- Clear instructions on use of water-resistant drywall.

Trim Carpenters

Working near the end of the construction process, trim carpenters and painters fix all the preceding trade's mistakes and are the cosmetic specialists. Poor quality work by the other trades significantly affects the trim carpenter's efficiency. Good quality framing, mechanical work, and drywall makes the trim carpenter's job easier. Trim carpenters normally need the interior of the house to themselves during their phase of construction. Other things that help them are as follows:

- A clean work area, all drywall mud off window frames, etc.

- Rough opening framed correctly, widths and heights correct measurements, square and plumb.

- All stairway trim and detail (including door swings) documented in drawing or plans.

- Coordination with floor covering contractors on base and shoe molding installation.

- Temporary heat during winter construction.

- Base, casing, chair rail, and crown molding details called out on plans.

- All materials available the morning of or day before trim carpenters' arrival on site. The most common problem reported by trim carpenters is having to wait for or chase after needed materials.

- Details on shelf locations and sizes.

- Backing for bath hardware.

Painters

- All color selections communicated to painter well in advance of paint start day.

- Site to themselves.

- Dust-free windows, doors, trim, and floor.

Floor Covering Installers

- Floor areas level and free from large drywall, mud droppings, etc.

- Transition lines between floor coverings (carpet, vinyl, tile, hardwood) clearly marked on plans and floor.

- Baseboard correct height.

The other trades have similar requirements specific to their phases of the construction process. The construction manager or superintendent should be trained to understand the concept of comfort zones and the benefits associated with keeping a trade within his comfort zone. Beyond the time efficiencies built into this approach, a trade will be much more likely to extend a favor to a construction manager in a bind if that same manager normally keeps him comfortable. When a trade is working within his comfort zone, he or she can generally produce more work in less time. This provides the trade and the builder with important advantages over the competition. The trade can complete more jobs and therefore make more money. As a result, the builder can get better pricing from the trade contractor. Additionally, when committed to be in two places at one time (don't they do that where you do business?), the trade will come or send the bulk of the crew to the job site where he or she will be the most comfortable, be able to complete the job, and bill the fastest.

The trade who can see the long-term benefits associated with working for the builder who "takes care of him" is the trade who is motivated to do timely and quality work. Next to prompt payment, this is the most important ingredient in motivating trade contractors.

Planning/Partnering with Trade Contractors

Successful application of the elastic comfort zone concept does require some specific training measures by the construction manager or superintendent. The trade should be brought into the planning process as the manager strives to keep the trade in his or her comfort zone. During this process, the builder and the trade can agree on the scope of work. This agreement and its enforcement allow the subsequent trade to start in his or her comfort zone. This "shared planning" is an indirect form of training and is the type most successful with the majority of trade contractors. At first, this experience will take the trade out of his comfort zone. It may be the first time that he has been asked to participate in such a meeting. The builder must do all he can to ensure a comfortable and positive atmosphere exists. As the trade becomes comfortable with an open two-way communication, his comfort zone, being elastic, will expand and the first step of the training process will have been accomplished.

By participating in the planning process, the trade will be more receptive to suggestions and will also suggest things, becoming an active participant in the planning process. This will make the plan his plan as well as the builder's; there is immense motivation associated with working to "your" plan as opposed to "their" plan. Complaining about unfair and "stupid" practices and procedures is easy when they are someone else's. In this vein, trades are some of the biggest complainers. Ever wonder why? When procedures and practices are part of their own plan, the complaints and gripes turn into suggestions for improvement.

During this planning process, the trade and the builder openly communicate. This enables a long-term, continuing dialogue to become part of the comfort zone, allowing the builder and the trade to constantly monitor and improve their working relationship. This sets the stage for training and learning.

Most builders fail in attempts to train trades because they take an approach that is too direct. Such an approach includes the traditional "why don't you do it this way?" and the similar "I'm the builder, therefore, I'm the boss" scenarios. Most trades are resistant to this approach. They are entrepreneurial in nature and often think they know all that they need to know because of their technical abilities. Such training attempts abruptly take most trades out of their comfort zone. The builder's responsibility is to use an approach that improves the communication, customer service, general business understanding, and, eventually, the performance quality of the trades.

Several avenues of direct trade contractor training exist that may be useful. Inviting a group of trades to a seminar-type setting is often less threatening. Many moderate to large builders have successfully combined such a session on safety, quality, customer service, or an introduction to a new plan or line of plans with a luncheon or information sharing opportunity. These sessions have to be well planned and convenient for good attendance. The time should be limited to a couple of hours. Holding such sessions on a weekend or in the evening helps to increase attendance. In addition, distributing checks after the meeting helps attendance greatly.

Other builders have had success in providing trade contractor manuals, which are distributed to each new trade contractor (see Figure 3.1). This manual explains company procedures and policies. It can include a good degree of operational policy training and even some technical training if each trade has a section in the manual. Inclusion of production manuals, contracts, specs, standard details, and quality checklists may make the trade contractor manual more useful.

Another approach is the recurring trade contractor newsletter. Produced periodically and distributed to the trades through the mail, these newsletters can keep the trades informed of company events and activities and can include training issues and success stories. "Trade contractor of the month" (or quarter) is often a popular feature and a source of some competition and pride within the trade base. The availability of computerized desktop publishing software allows for the creation of professional newsletters at a minimum of expense.

Motivation and training can be enhanced by the use of "I caught you" cards or coupons. These cards or coupons are given by construction managers to trades who have been "caught" going the extra mile or doing a consistently good job, etc. Collecting a certain number of these cards or coupons allows the trade to redeem them for gifts or dinner/entertainment tickets. Builders of all sizes have used these with success.

When the builder takes the time to communicate, encourage, and reward trade contractors (beyond the contract payment), most will respond with increased quality and operational efficiency where this builder is concerned.

FIGURE 3.1 Sample Page from a Trade Contractor Manual

My Building Company's
Trade Contractor Manual

Table of Contents

Comfort Zones Require Two-Way Communication

So far this chapter has concentrated on the methods of keeping and training trade contractors within their comfort zones. This is a two-way communication process. Not only do builders and the trades have to discuss and reach agreement on things that will keep the trade in his comfort zone but everyone must exercise the win/win philosophy. The builder, throughout the dialogue, should search for opportunities for mutual operational improvement. For example, as the builder is agreeing with the excavator to stake the desired over-dig, as well as the exact house dimensions, it may be a good time to negotiate that henceforth the excavator will slope all trenches over 18" in depth to prevent cave-in and to comply with the OSHA regulations.

What Trades Can Offer In Return

The following is a partial list of potential points that builders should include in their trade contractor agreements to ensure a mutual comfort zone. These may be included in terms, agreements and/or scope of work contract sections.

- Prompt response (with full crew) to scheduled work when proper lead times are given.

- A professional approach to communications with construction managers and customers under warranty. You should be able to work together to develop a mutually convenient method of communication for even the smallest of trades.

- Respect for agreed upon payment terms.

- Professional appearance and job site demeanor. Loud music in subdivisions or populated areas should be avoided. Use of alcoholic beverages and/or drugs is forbidden. A professional approach to customer communications at the job site is an important part of the job site demeanor.

- Job site cleanliness should be every trade's responsibility. Use dumpsters and other trash receptacles, broom sweep the site at the end of every shift, etc. Not using the tubs or sump pump crocks as urinals goes without saying. Nothing sets the desired atmosphere of efficiency as quickly as a clean, well-organized job site.

- The builder/trade relationship should evolve to the point where both parties understand that their mutual livelihood depends upon the builder's reputation and credibility. The trade and the trade's employees should act accordingly. Biting the hand that feeds you is an all too common occurrence among trade contractors.

Additionally trades who are well treated and who are working in their comfort zones should be more willing to work out the scope of work details so vital to a smooth construction process. Examples of these details include: Who installs the roof felt paper: the carpenter or the roofer? How about the flashing on the chimney? The insulation and green board behind the bathtubs and shower stalls? Who installs those tubs? Ways to resolve these issues should be well coordinated within a builder's operation and not left to the construction manager to handle. Everyone should be kept in their comfort zones by negotiating mutually reasonable solutions. Two-way communications will produce the results by addressing the little details that can add up to big headaches.

4

Negotiating and Pricing

Afters finding trade contractors who can do the job, an important step is yet to come. You must establish a firm and fair price and agree upon a scope of work, which can be documented in the trade contractor agreement. This is called negotiating and is a two-sided process. During negotiations, keep the win/win philosophy in mind. A one-sided negotiation eventually leads to major problems for all involved. Even the party who feels they have "won" at the negotiation table eventually suffers from disillusionment on the part of the other party. In win/win negotiations, the major goal of both parties should be a mutually profitable *long-term* relationship.

In builder/trade contractor negotiations, price, scope of work, and timeliness of communication and construction duration should all be discussed and agreed upon.

Price

Based upon the scope of work, a firm price should be agreed upon. A firm price is one that is locked in for an

Production builders or semi-custom builders should try to lock labor prices in for a specific period of time. A 3- to 6-month lock is most common. Labor price locks of up to a year are often possible and are always preferred. These price locks are crucial to the success of the builder. It is hard to sell homes profitably at today's prices for construction in 3 to 6 months without price locks at least on the labor side.

agreed upon period of time or number of jobs. If you bid each job, the firm price is the one for that job.

When negotiating pricing, remember to agree upon prices for all scope of work components that are priced separately. For example, the brick mason's normal pricing consists of the labor rate for laying one thousand bricks. Does this price change from one-story to two-story homes? What does the mason charge to place quoins on the corners of your upscale homes? How about placing keystones above windows, soldier courses above garage

A Framing Pricing Technique to Consider

Rough frame carpenters in many parts of the country price their labor to builders and general contractors on a square footage basis. This is a simple and easy to calculate formula and therefore remains quite popular. However, a major downside to both framer and builder is that the square foot framing labor pricing formula does not allow for pricing differences between simple or difficult to build homes. There is no price break for the builder who used roof and floor trusses versus stick framing. There is no difference for a 5/12 roof pitch versus a 10/12 pitch.

There is no difference for the 2,000-square-foot home with 400 linear feet of partition walls and one with 650 feet of interior wall. There is no difference for slab or crawl space or basement construction. Basements and crawls requiring another floor system should cost more. Slabs should cost less.

A comprehensive potential solution is to price out each component of framing. Price the square feet of roof sheathing to be applied. This will make up for the roof pitch variations. Price the square feet of floor decking. This will make up for the differences in floor areas for different foundation types. Have a price difference negotiated for use of trusses versus stick framed roofs. Price the linear feet of partition walls, a price per window and door opening as well as dormers, covered porches, and basement walls.

This system could become complex and overwhelming. You might want to start out with a hybrid system that takes into consideration the complexity of the structure being built, not just the square footage.

door, lintels or bricking around decorative vents? Is there a charge for placing address stones? Is the rate the same for fireplace chimneys where additional scaffolding must be used?

A builder needs to know this detailed pricing information to make correct sales price quotes to customers. This is especially true in remodeling projects where there is less room for pricing errors. The estimating process should include procedures to calculate costs for each scope of work item that can be priced separately.

Confirming You Are Not Paying Too Much

Even small volume builders need to get competitive pricing information on a periodic basis. You may use only one source of concrete labor, but you will want to price concrete labor on a semi-annual basis. If you use a single source of labor and never check alternate pricing, you may soon find yourself paying too much.

Maintain a competitive pricing chart complete with detailed scope of work price break-outs and the date of each price quote. Figure 4.1 shows a representative competitive price quote chart. Compiling this chart takes some time in interviewing potential trades. In the worst case scenario, the piece of mind in knowing that you are still paying a competitive rate is worth the effort. In the best case, you will discover some negotiating ammunition to help keep your prices in line.

However, price is not the sole ingredient of a successfully negotiated labor contract. Quality, scope of work, timeliness of construction start, duration, clean up, and payment terms should all be equally considered.

Scope of Work

When negotiating, a full understanding of what the builder expects is crucial for the trade contractor. Without this information, pricing will be incomplete and many field level misunderstandings may result. These misunderstandings will slow construction and cost everyone valuable time and money.

A list of duties, or scope of work, to be performed should be developed for each trade contract. It should be very detailed and include discussion of type of materials and installation methods.

Some Common Scope of Work Questions

To stimulate the reader's thinking and to further describe the detailed scope of work, let's look at some common questions within a home builder's scope of work:

FIGURE 4.1 Comparison Price Quote Chart

Vista South Custom Homes
Vendor Price Comparison Report

Trade:_____

Date: _____

Scope of Work Resource	Unit	Berry	Scotts	Rayburn
6/12 Pitch 20 year	Square	19.50	25.00	16.00
8/12 Pitch 20 year	Square	26.00	32.00	16.00
9/12 Pitch 20 year	Square	29.25	36.00	18.00
10/12 Pitch 20 year	Square	32.50	40.00	20.00
11/12 Pitch 20 year	Square	35.75	42.00	22.00
12/12 Pitch 20 year	Square	39.00	44.00	24.00
Second Story Charge	Square	5.00	—	—
Flashing—Step	Ln. Foot	3.00	2.50	—
Flashing—J Channel	Ln. Foot	3.60	2.50	—
Ridge Vent	Ln. Foot	2.00	2.50	—
Drip Edge	Ln. Foot	1.10	1.00	1.00
Chimney Flash	Each	by foot	175.00	—
Upgrade Dimensional	Square	10.00	9.00	4.00
Skylight 2×2 or 2×4	Each	125.00	100.00	75.00

- Who installs the roof drip edge and/or roof felt paper?
- Does the builder expect the drywall to be screwed or nailed on interior walls?
- Who supplies the flashing for roofs and chimneys—builder or trade?
- Does the builder expect the subfloor to be glued and screwed to the floor joists or trusses?
- Who installs the underlayment under vinyl flooring or hardwood?
- Who installs shoe molding around cabinets—flooring or trim contractor?

- Does the builder expect the basement excavator to dig a sump pit external to the basement dig?

- How does the builder expect his corners and wall intersections framed?

- Is 54"-wide drywall used in rooms with 9' ceiling heights?

- Who digs the plumbing supply and waste lines and when?

- Do you allow staples to be used during roof shingle installation?

These and many other questions specific to the way you do business should be answered in the scope of work section of the builder/trade contractor negotiation.

Timeliness of Service

The third negotiable ingredient relates to time. To be most efficient, both trade and general contractor need to know each other's time requirements. How long before the trade is to start work does he need notification? This is called lead time. How long will the trade's phase of construction take? This is called duration. How long will the builder wait for warranty service to be scheduled and performed? This is called service schedule.

The lead times, durations, and service schedules need to be mutually agreed upon to ensure everyone stays within their comfort zones. Construction schedules are dependent upon duration. Staying on schedule depends upon adhering to lead times, and customer satisfaction is often dependent upon following the service schedule.

Time is the least understood ingredient of construction success. Cost or price and quality of workmanship are more easily related to success. Time is equally important. Over the years, I have seen more dissatisfied customers as a result of not adhering to the original schedule than all other causes combined.

Reach and document agreement with each trade on lead times. The builder needs to notify the trade x number of days before construction starts to get on the trade's schedule. The trade needs to notify the builder if he is unable to keep the original schedule x number of days prior to this original start date.

The trades in my market that require the longest lead times are the roofers and the framers, who often require 30 days. We then follow up with them 10 days prior to original start date and confirm or change the anticipated date. The other trades all require less than 30 days. The required lead

Roofer Asks for 30-Day Lead Time

Our roofing contractor has recently switched to a 30-day lead time. This literally requires us, as the builder, to schedule the roofer before the foundation is dug. The roofer then tracks the project's progress on his daily trips around town and adjusts his schedule to be on our site as needed. He is proactively working his schedule. He says that scheduling problems cost him $40,000 last year and he can't afford another year like that.

Our roofer runs three crews, and it takes him about 15 hours per week to work this schedule in the proactive fashion. He estimates that this operational change will result in him having record sales volume with less inactive labor hours, and he may even be able to work the increased volume with two crews. **It will probably result in an $80,000 profit swing in one year.**

As the builder, his lead time need requires us to be disciplined and produce purchase orders well prior to construction start. After this, the roofer's proactive stance on scheduling makes our job easier. Sometimes we call and say we will be ready next Tuesday and the roofer already has it scheduled for that day. Partnering does work.

time is common to any group of trades, based upon the business volume and business savvy of the contractor.

Construction Duration and Crew Size

How long will this job take? Each residential construction project is different. However, there are some standard durations that can be estimated when using standard specifications of labor and material. How long will it take to frame a 2,000 square foot two-story home? There should be a negotiated standard duration for this event. My framer and I have negotiated a five full-day duration for this type of home. For me, this information is vital in scheduling the follow-up trade contractors (i.e., rough mechanicals, inspection, and roof). It is also vital to the framer so that he can schedule his next job and give the next general contractor on his list an accurate start date.

Often, to meet the agreed upon schedule, a trade contractor will need a crew of a certain size. This is important training for most trades to learn the impact of crew size and level of experience on duration. I have worked with many trades who have formed a second crew only to find the second crew's durations were much longer than the first crew's. The experienced crew number one had a synergy that was not easy to replace when the crew was split to form the second crew.

Durations must be flexible because of weather variations and trade crises. We allow a plus or minus 20% factor in our overall construction schedule for these impacts. Thus, a 5-day frame job would be scheduled for an extra day or 6 days total in our master schedule.

Price, scope of work, and schedule should be reviewed and negotiated as equal ingredients to the builder's and trade contractor's partnering success. They should be constant points of discussion. If they need to be adjusted, then a mutually accepted compromise is normally the best route. It is far better to spend extra time up front and get all three negotiation points resolved to the best of both parties' abilities. Start with a workable plan and try to stick with it.

5

Scheduling Trade Contractors

Keeping trades, suppliers, and employees in their comfort zones takes a great deal of planning. A large part of any construction planning process is scheduling. Scheduling requires planning the sequence and duration of all the construction activities, including the construction items, material ordering and delivery, color and decorating selections, etc. One of the most important ingredients in the scheduling process is an understanding of the interdependencies of the trade contractors on the job site(s).

Jerry Householder, in his book *Scheduling for Builders* (NAHB, 1996), hails formalized scheduling methods as a major breakthrough for the construction industry. He also points out that residential builders have not, as a whole, taken advantage of this management tool. Householder adds that many builders manage by reacting to one "crisis" after another, putting out fires as they arise on their job sites. This is an expensive and exhausting way to operate. Formalized scheduling techniques allow you increased control of the work and the finances. These techniques also help the builder maintain good relationships with all trade contractors and suppliers.

Within the builder/trade contractor relationship, the most complex issue is ensuring that the construction schedule is mutually understood and accomplished. This chapter explores methods of easing the scheduling process and maintaining the established schedule.

The Process of Scheduling

Some of the basics of scheduling must be reviewed to ensure a proper understanding of this chapter's contents. Successful construction scheduling requires the following:

- **A list of all construction activities in the job**—Subdivision builders should produce the activity list for each model and then overlap the unit lists for all the construction within a given period for the project. The activity list should include the steps in the construction process and items such as inspections to be scheduled, orders to place, color selections, etc.

- **A duration calculation (in work days) for each activity**—The duration should be specific to the trade contractor as well as to the trade. The builder should not assume each framing crew will maintain the same schedule. If the builder uses multiple contractors within the same trade, it will be necessary to adjust the base schedule to the specific contractor once an assignment is made. Durations are often hard to define without experience with a particular trade. Noted production management consultant Leon Rogers of Brigham Young University says that the builder should keep records of current job activity, which will allow historical application of the rational method of duration calculation. The rational method measures the units of construction work (number of block, yards of concrete, square foot of frame, etc.). Then the method calls for measuring past performance of the units of work on a per crew member basis. This determines productivity rate per crew member. Then with future trades a starting point would be to divide the units of work by the productivity rate.

- **Scheduled advance lead times**—Lead times for labor and material deliveries must be compiled. These lead times should be backed into the schedule after scheduling the needed construction activities and delivery dates. This is the most often overlooked area of formalized scheduling as well as the area providing the most opportunity for positive builder/trade contractor team building.

- **The activities list in proper sequence**—Some activities must precede other activities. For example, the mechanical inspection must be complete before the insulation can begin. However, other activities can happen simultaneously, such as finishing electric and trim carpentry. Some activities are subject to sequence changes from builder to builder and area to area, for example, concrete flatwork. The simplest method of illustrating a sequenced schedule is with the use of a bar chart type schedule. Figure 5.1 shows an example of a simple bar chart schedule.

- **A critical path**—Within any construction schedule, a "critical path" surfaces. The critical path includes activities whose flow cannot be changed sequence-wise. Changing the duration of any of the critical activities will cause the greatest disruption. The critical path emerges most clearly in precedence network or flow chart type schedules. See Figure 5.2 for an example. Neither Figure 5.1 or 5.2 show the additionally needed trade contractor advance notices as part of the activities schedule. The author recommends adding these advance notices as ticklers to the scheduling system. This allows the construction manager to effectively communicate to the trade and allows the trade to plan his workload at least several weeks in advance.

With the activity list, the duration calculations, and the advance notice or lead time lists in hand, the schedule can be developed. Principal trade contractors can have a hand in the scheduling process by providing input on durations and sequence of construction and by identifying the other trades that they can and cannot overlap with on the job site. They also should have a full say in establishing their desired lead or notification times. This input, when received up-front, can go a long way toward avoiding conflict on the job during construction.

A Successful Scheduling System

The FAST Management Group teaches one of the most thoroughly documented scheduling processes seminars entitled "Perpetual Scheduling." The FAST Management Group, from Redmond, Washington, is a developer of an integrated software program for home builders. They specialize in assisting large-volume builders in scheduling, purchasing, estimating, accounting, and most other industry related systems. In fact, their FAST acronym actually stands for "Fixed Activity Scheduling Technique." Their perpetual scheduling system theories are equally efficient for any size builder.

The perpetual scheduling system calls for the use of the following techniques:

- Develop your current activity list and sequence for one-house plan. Include only activities like calls for inspections and advance notice to trade contractors.

- Write out a complete critical path method (CPM) flowchart.

- Refine sequences, durations, and critical activities with the help of employees, trades, and inspectors.

- Computerize your original schedule and distribute a hard copy (printout) to your team (including trade contractors) for review and concurrence. If you do not have a computer, photocopy the schedule for distribution to the team. Then send someone out to research your computer needs and to buy a computer for your operation.

- Review and edit the schedule with the entire team.

- Follow and use this schedule for all of your plans within a project. A project could be a trade division or a geographic area where units under construction share a common trade contractor and supplier base.

- Use a superintendent or construction manager who will support formalized scheduling procedures.

- Distribute 2-week production schedule reports that detail the upcoming scheduled activities on each house.

- Merge the schedules of the houses under construction to look for trade contractor and superintendent overload. If necessary, adjust the schedules to compensate for the resource overload.

- Update your schedule whenever something changes. In the beginning, this may be daily until the schedule gets "tweaked" and the team learns to "work the schedule." Adjusted schedules should be produced as necessary. This is why scheduling is much easier with a computer. The edits and new report printing are easier.

- Consider hiring or assigning an individual to be the schedule coordinator in larger companies. This person would be the communica-

FIGURE 5.1 Bar Chart Schedule

		Month 1				Month 2	
Week Numbers		Week 1	Week 2	Week 3	Week 4	Week 5	Week 6
Activity	Duration Days						
Project Setup	4 days	▤					
Sitework	10 days		▤	▤			
Foundation Work	10 days			▤	▤		
Install Footings	4 days			▤			
Set Batter Boards	1 day			▤			
Excavate Footings	2 days			▤			
Set Grade Stakes	1 day			▤			
Footing Inspections	1 day			▤			
Place Footing Concrete	1 day				▤		
Install Foundation Block	3 days				▤		
Deliver Block	1 day				▤		
Lay Foundation Block	3 days				▤		
Slab On-grade Work	10 days					▤	▤
Utility Rough-in	3 days					▤	
Plumbing Rough-in	2 days					▤	
HVAC Rough-in	1 day					▤	
Dryer Duct Rough-in	1 day					▤	
Slab Preparation Work	4 days					▤	
Fine Grade Dirt	2 days					▤	
Spread Gravel	1 day					▤	
Termite Treatment	1 day						▤
Place Poly and WWM	1 day						▤
Set Grade Stakes	1 day						▤
Place Slab	1 day						▤
Inspect Slab	1 day						▤
Place Concrete Slab	1 day						▤
Structure Phase	20 days						
Erect Rough Framing	15 days						
Deliver Framing Material	1 day						▤
Wall and Sheathing	10 days						
Roof Framing and Deck	5 days						
Finish Framing	5 days						
Check Plumb and Square	2 days						
Check Door Opening	2 days						
Install Deadwood and Blocking	2 days						
Utility Rough-in	8 days						
Plumbing Top-out	3 days						
Plumbing Test & Inspection	1 day						
HVAC Rough-in	3 days						
Electrical Rough-in	2 days						
Electrical Inspection	1 day						
Final Rough Framing Items	2 days						
Wall Insulation	1 day						
Framing Inspection	1 day						
Clean Up All Waste Material	1 day						
Structure Dry-in Phase	7 days						
Finish Phase	20 days						
Project Close-out	5 days						
Final Purch List	1 day						
Final Inspection	1 day						

Source: Reprinted from Thomas A. Love, Figure 5. "Chart Showing Overall Schedule," *Bar Chart Scheduling for Residential Construction* (Washington, DC: HBP, NAHB, 1997) p. 6.

FIGURE 5.1 Bar Chart Schedule *Continued*

		Month 3				Month 4				Month 5
Week 7	Week 8	Week 9	Week 10	Week 11	Week 12	Week 13	Week 14	Week 15	Week 16	Week 17

FIGURE 5.2 Precedence Network

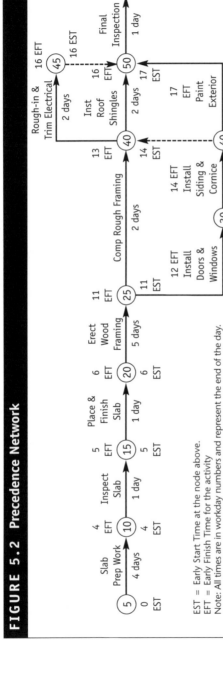

EST = Early Start Time at the node above.
EFT = Early Finish Time for the activity
Note: All times are in workday numbers and represent the end of the day.

The chart below shows the information calculated during the forward pass.

Activity Description	Early Start Time	+	Duration	=	Early Finish Time
Slab Prep Work	0	+	4	=	4
Inspect Slab	4	+	1	=	5
Place & Finish Slab	5	+	1	=	6
Erect Wood Framing	6	+	5	=	11
Comp Rough Framing	11	+	2	=	13
Inst Doors & Windows	11	+	1	=	12
Inst Siding & Cornice	12	+	2	=	14
Install Shingles	14	+	2	=	16
Paint Exterior	14	+	3	=	17
Rough-in & Trim Electrical	14	+	2	=	16*
Final Inspection	17	+	1	=	18*

* The EST for Rough-in & Trim Electrical follows the critical activity, Install Siding & Cornice, and therefore reflects the preceding critical activities with the longest durations. Similarly, the Final Inspection cannot begin until all preceding critical activities have been completed.

Source: Reprinted from Thomas A. Love, Figure 2.11. "Forward Pass Calculations," *Scheduling Residential Construction for Builders and Remodelers* (Washington, DC: HPB, NAHB, 1995) p. 21.

tions point between the superintendent(s) and the trade contractors, suppliers, inspectors, and possibly, the customers. The schedule coordinator would also update the schedules on a daily basis and ensure that all interested parties have copies.

- The coordinator's largest responsibility would be to initiate a scheduled work order confirmation procedure, which he reviews with the construction manager(s) each morning (or evening for the next day). Then the coordinator makes calls to the trades, suppliers, and inspection departments to confirm any scheduled work, delivery, or inspection. Next, the coordinator reports exceptions, problems, or decisions to the construction manager or superintendent. Finally, schedule variances are entered into the scheduling system and updated schedules are produced.

In the perpetual scheduling system, the schedule coordinator must be reliable, detail oriented, personally organized, efficient, and have a good telephone manner. The coordinator must also have a basic knowledge of construction sequence and terminology. This system, including the coordinator position, is being used successfully by several moderate- to large-volume residential builders around the country. The perpetual scheduling system's largest benefits include improved communication and accurate expectation levels between builder and trade contractor. Many thanks to FAST for sharing their system's success formula.

Keys to Scheduling Trades

The nine key rules for successfully scheduling trade contract labor are as follows:

- Always have two-way "lead times" built into trade contract agreements. Two-way means an agreed upon advance notice for the trade indicating when the builder needs the work to begin and an agreed time for the trade to let the builder know of conflicts and problems.

- Explain your expectations. During trade contract negotiations explain your organization's commitment to communications and adherence to the schedule. Let the trade know that the first way to ruin your confidence is failure to show as scheduled or failure to inform you of delays and problems.

■ Agree upon a mutually convenient method of communication by which the trade contractor's work will be scheduled (for example, paged messages or calls will be returned within one hour). With today's modern devices, communications should not be a problem, yet it remains one of the largest problems faced by most builders. Pagers (both numeric and alphanumeric), cellular phones, fax machines, and e-mail have all been added to the tools available to the average trade contractor in the last decade. What we need to do is agree on how and when these tools can be used. There should be little need to make calls at 5 am or 10 pm as we had to do five to ten years ago to talk to our trade contractors.

■ Schedule advance notice communications as part of your construction schedule/plan.

■ Two days before a trade is scheduled to begin work on your site, call to confirm the intention to start on time. Never wait for a trade contractor to call you about a schedule conflict. If you wait, you will definitely have a frustrating experience.

■ Discuss crew size during negotiations and have field personnel monitor the trade's adherence to the agreed upon crew size. A trade is becoming over-committed when he starts splitting his crews. It is better for him to tell you that he can't start for another three days than for him to take six days for a three-day job. You can often find a replacement trade or move another phase ahead. You will be in control more than if you let the trade expand his duration without any proactive action on your part.

■ Monitor your trades and your schedule. Develop a system that allows you to track schedule variances and the party responsible for the delay. Details of such a system are offered in the next section of this chapter.

■ Utilize the results of the schedule variance monitoring to evaluate the efficiency of your trades. Remember that price, quality, and TIMELINESS OF WORK (including showing up as scheduled) should all contribute to the evaluation of the trade contractor's performance.

■ Never get complacent; always look for improvements in your trade contractor base. Constantly interview new trades and look for training opportunities with existing trades.

Preconstruction Meetings

For years, many commercial builders have enjoyed great benefits of conducting formal preconstruction meetings. Few residential builders have applied this management technique to their operations. The advocates of these meetings testify to the positive influence such meetings have upon their construction schedules. How might these meetings apply in a range of residential construction operations?

Who Attends?

In the commercial environment, attendees typically include representatives from the trade contracting firms, project engineers, material suppliers, construction managers, property owner/developers, financing or banking participants, and the municipal building department. These commercial projects are generally large scale and complex with everyone involved having a major interest in attending preconstruction meetings.

In the residential environment, this is not usually the case. The residential builder must experiment to determine if this type of meeting can be adapted to his operation. Often two meetings, one with the presale customer and one with trades, construction managers, and sometimes suppliers and inspectors may be appropriate. Custom home builders often find this a useful opportunity to set expectation levels, to discuss the complexities of the home in question, and to review all aspects of the schedule. Production builders may find that preconstruction meetings are appropriate at the opening of a new work phase within a community or subdivision. Production builders may want other production employees, such as estimators or schedulers, involved in their preconstruction meetings.

Smaller scale semi-custom builders benefit by periodic (quarterly, semi-annual, etc.) construction meetings with their team of suppliers, inspectors, and trades. The semi-custom builder should also have a construction-specific meeting with each homebuyer to set mutual expectation levels.

What Happens?

Common to all these types of "preconstruction" meetings are several ingredients. All attendees should receive an agenda, plans or prints, and a copy of the construction schedule far enough in advance to do any necessary review and preparation. Agendas should include the date, time (both beginning and ending time), and place of the meeting, as well as a list of items, if any, that the participants need to bring.

The meeting is an opportunity to communicate. As such, minutes should be taken, a summary typed, and copies distributed as soon as possible after the meeting. The summary meeting report should detail all decisions and commitments. The reports should be forwarded to each participant with a cover memo requesting that any inconsistencies and inaccuracies be brought to the builder's attention immediately for correction.

During the meeting, the builder wants commitments from trades on durations, notice lead times, etc. He should also review decisions to be made, summarize the decisions that are made, and solicit questions and feedback on stated plans. The builder should stick to the agenda and start and end on time. If attendees are to agree to attending future meetings, then beginning and ending on time is crucial. Do not wait for someone who is late. Let the late party stay after the meeting to review what they missed.

Meetings should be held at an office where interruptions will not be possible, with someone screening calls and visitors to prevent interruptions. This type of meeting should never be held on the job site. With the distractions and interruptions, agendas will quickly crumble. If no office is available, try to rent a private room at a restaurant, a hotel, or in a public building such as a school, library, or municipal building.

Progress Meetings

For many larger projects, a single preconstruction meeting will not be enough. These large projects often require follow-up meetings, where the plans and decisions are monitored and schedules and budgets are often adjusted. These meetings should include standard progress reports such as schedule updates, cost reports, and engineering reports.

Large projects, including trade division phases, should also have a post-construction review meeting. This meeting will give the parties a chance to discuss the successes and failures of the project. This will serve as an excellent planning tool for the next project undertaken. Hold this meeting as soon as possible after the completion of a project while all the details are fresh in everyone's mind. Again, have minutes taken and distributed. These minutes should be used at the preconstruction meeting for the next project.

Weekly Production Meetings

Many large volume builders find it advantageous to hold weekly production staff meetings. These meetings are attended by all construction employees and as many trade contractors as is practical. The construction schedule as a whole is a major agenda item. Priorities are set, job specifics are discussed, and trade interactions are scheduled. Often a master subdivision or project schedule is developed for the next several weeks. Figure 5.3 shows a sample two-week vendor-specific schedule.

Monitoring Your Construction Schedule

Figure 5.4 shows an example page from a schedule variance tracking system developed for use by residential builders. This schedule variance report allows the builder to monitor actual construction progress against the projected schedule. A clerical staff person (perhaps the production coordinator, as discussed in the previous section) can maintain this report. It can serve as a planning tool for superintendents and as a monitoring tool for company owners and/or upper management. It can help the builder realize training opportunities within his tradebase.

The report breaks the construction cycle down into six phases (your system may adapt to different phases) and assigns a time budget or projected schedule for each construction phase and for the unit as a whole. The report then compares the actual schedule to the projected schedule and identifies (through the use of variance codes) the cause of any delay.

Tracking this information and monitoring for trends will often help identify problems causing expensive schedule runs. You will be able to pinpoint which phase of construction causes the most delay, which trades contribute to schedule variances, and even monitor the specific reasons for the schedule delays. These reports can provide a common ground for discussions

---

FIGURE 5.3 Two-Week Vendor-Specific Schedule

Vendor Loading Schedule   Division: _20 BV_

Vendor: _RS Wilson Plumbing_   Code: _rsw_   Date: _____

Activity/Job #	mon 8/1	tue 8/2	wed 8/3	thu 8/4	fri 8/5	mon 8/8	tue 8/9	wed 8/10	thu 8/11	fri 8/12
Finish 1008	▨									
Rough 1012		▨	▨							
Water Line 1012		▨								
Rough 1015					▨	▨				
Slab 1015							▨	▨		
Inspection 1018										▨

FIGURE 5.4 Schedule Variance Report

Schedule Variance Report
Date: September 10, 2000

Job # 201	Phase	Goal Start Date	Actual Start Date	Goal # Days	Actual # Days	Variance # Days	#1 Variance Code	#1 Contributor	#2 Variance Code	#2 Contributor
Model	1	May 1	May 2	10	9	−1	C			
SL1	2	May 11	May 10	15	18	3	D1	Carpentry Co.		
	3	May 26	May 28	15	20	5	A	X County		
	4	June 12	June 19	20	16	−4				
Supt.	5	July 1	July 5	25	23	−2				
TS	6	July 26	July 28	5	5	0				
Total		Aug 1	Aug 2	90	91	1				

Job # 202	Phase	Goal Start Date	Actual Start Date	Goal # Days	Actual # Days	Variance # Days	#1 Variance Code	#1 Contributor	#2 Variance Code	#2 Contributor
Model	1	June 1	June 1	10	12	2	G1	Soil Cond.		
R2	2	June 11	June 13	15	18	3	D1	Carpentry Co.	F3	Lumber Co.
	3	June 26	July 1	17	18	1	D4	Plumbing Co.	G2	Staff
	4	July 15	July 19	20	18	−2		Drywall Co.		
Supt.	5	Aug 5	Aug 7	25	30	5	D1	Plumbing Co.	H1	Molding Change
TS	6	Aug 30	Sept 7	5	6	1	G2	Staff		
Total		Sept 4	Sept 14	92	102	10				

Phase numbers:

1 Dig through foundation
2 Foundation through frame
3 Frame through mech and insulation
4 Insulation through drywall finish
5 Drywall through walk-thru
6 Walk-thru to closing (completion)

Schedule Variance Codes:

A Inspection-code related

B Weather related

C Improper prints/plans

D Trade contractor related
  1—Did not show as scheduled
  2—Quality correction problem
  3—Inadequate trade base
  4—Scheduled time overrun

F Material related
  1—Incorrect material sent
  2—Incorrect quantity sent
  3—Damaged material sent
  4—Theft
  5—Vandalism

G Superintendent related
  1—Adverse site conditions
  2—Improper lead time to trades
  3—Coordination trade and material
  4—Management decision
  5—Prior phase not complete

H Customer caused delay
  1—Change order
  2—Selection delay

I Other

with trade contractors. A builder can be more specific when addressing where and how a trade's performance or lack of it creates problems for the team. When a document, like this report, is the focus of such discussions the builder has a tool that allows for objectivity and avoidance of a personal affront.

The schedule variance report can be updated and printed as needed. Periodically, this information can be summarized into an additional report that allows tracking of how many variances of each type occurred and how many times a particular trade contractor or supplier contributed to the schedule variances. Monitoring these reports on a consistent and regular basis provides the basis for improved business decisions.

6

Trade Contractors: Quality Control

Along with the labor shortage, degradation of construction and general work quality are the main problems facing residential contractors today. The old craftsman work ethic is very difficult to find. Pride in workmanship is not respected as it once was. Everyone talks about quality, but today quality seems to refer to operational efficiency and customer satisfaction, not to pride in, individually, doing a good and thorough job.

Maybe it is the "what's in it for me" attitude or the hectic pace caused by the production schedule meaning more than the individual worker's sense of accomplishment. For whatever reason, the trend continues to worsen each year.

The problem developed as the apprenticeship programs for the trades declined. Maybe the technical training taught more than just how to perform a particular trade. Perhaps the training instilled a sense of pride in work well done, which converted into self-esteem. Recent

NAHB initiatives to re-establish this type of training program could go a long way toward improving our quality problems.

In a less theoretical vein, we can influence the final product quality on our job sites despite the general lack of individual pride in performance. It is much more difficult and time consuming than it used to be, but it can be accomplished. Most commonly, a systematic approach to checking and double checking each trade's performance must be followed to reach acceptable levels of product quality.

Another approach, often referred to as Total Quality Management (TQM), combines the checklist routine with an attempt to statistically influence worker performance and accountability. The Total Quality Management movement has, over the past ten years, gained acceptance in the light construction industry. TQM refers to a system of managing a process, which includes clearly defined goals, where all involved give input, and continual efforts are made to improve the process for all involved. TQM also has a monitoring component, which allows those involved in the process to track actual performance against a "benchmark" or predetermined goal. Statistical measurements are made and results documented against the goals to ensure a continually improving or consistent level of quality.

Like many of the current wave of management philosophies, TQM is founded in solid, common sense logic. In addition, like many of today's philosophies, it is humanistic in approach. The win/win philosophy discussed throughout this book is a byproduct of TQM.

Builders and trade contractors can apply many of the TQM components into their mutual management practices. It might be best, in most cases, to not dwell on the theory of TQM but jump right into some practices, which will provide recognizable impact with the trades.

A First Step to TQM

One of the easiest and most productive TQM steps involves getting the builder and trade contractor together to plan a project. The easiest way to do this is to have what I call a technical review meeting on a new plan or a large house or project you are getting ready to undertake.

In a technical review plan meeting, a set of construction plans on a new house or project is distributed to the key team members, including the trade contractors. It is especially important to have foundation, frame, and mechanical contractors attend. Each trade and team member participates in a thorough review of the details, measurements, and intricacies of the plan. Suggestions for improvement are normally plentiful. All team members leave with a renewed vigor for the potential of the project if their input has been listened to and considered. This is time, and often lunch money, well spent. This process will speed the actual construction process because the key team has already "built the plan in their collective head."

Good quality is the often-elusive ingredient in the construction success formula. Builders may contend that quality standards are hard to enforce with trade contracted labor. However, this contention contradicts one of the main reasons for hiring independent trade contractors, namely that unlike internal labor crews, trades are experts in specific areas of construction. Because quality is more difficult to measure than cost and time, builders often resign themselves to accepting undesirable quality levels or frequently changing trades in an effort to improve quality.

Work quality strongly influences the long-term success of most residential construction companies. Builders often feel that expense and quality go hand in hand. They believe that a high price must accompany good quality. In most cases, however, quality labor should not cost more than average labor prices. Labor costs depend on current market conditions more than on the quality of labor. Differences in labor quality depend more on the builder's varying standards of acceptability than on price. This chapter explains how builders can establish their own acceptable quality standards and then measure and maintain these standards before, during, and after the construction process.

What is Quality Construction?

The term *quality* is often used but rarely defined. *Quality*, in our industry, reflects the interplay of two independent ingredients: materials and workmanship. How do builders ensure the quality of a house that will stand the test of time and materials that will work well and be aesthetically pleasing? How can quality be defined so that it can be measured, monitored, and taught?

The answers lie in setting measurable quality standards. These standards should be reflected in the builder's materials specifications and in quality control checklists for labor and workmanship (see Appendices B and C). Construction managers and trade contractors should be taught how to use these checklists. These quality standards should become the foundation of any builder's production management system. They become an integral part of plan design, specifications, sales contracts, trade contractor agreements, walk-through checklist, warranties, and many other potential management tools.

Meeting quality standards becomes much easier when measurable standards are set and consistently enforced throughout an organization.

Management of construction quality is a vital part of the relationship between the builder and the trade contractor. Both parties must agree on specific quality standards and develop a structured method of ensuring that these standards are monitored and attained on each job.

Setting Quality Standards

This book concentrates on quality of work rather than quality of materials. Materials are a byproduct of design, and the architect or purchasing agent determines their specifications. Labor is heavily influenced by the trade contractor's relationship with the builder. Setting labor quality standards is a demanding task. You can begin developing quality standards by developing a formal company statement of quality (see Figure 6.1). It should be a general statement that makes evident the company's commitment to customer satisfaction and error-free services and products. All of your employees, including your trades, should be familiar with and uphold the contents of this statement. Many small items often are taken for granted and are not included in builder's quality standards. Quality control checklists can help ensure that all activities of each construction phase are completed in an acceptable fashion. You should develop checklists for each trade or phase of construction. These checklists should encompass and exceed local building code requirements.

All construction standards should be documented, often in a manual format, for distribution to all parties involved in the construction process. This will include employees and trade contractors. These documents should become part of the training programs and contract agreements for each team member.

FIGURE 6.1 Statement of Quality

Cooper Homes' "Statement of Quality"

At Cooper Homes we believe people should live better in retirement than before. To do this we must produce a product with zero defects and low maintenance. All involved in the process of building a Cooper home must be committed to the philosophy that...

"A Cooper Home is a reward for a lifetime of hard work."

Framing details, sheathing, subfloor, and drywall layouts can be included for all standard plans. You can find more information on format and how to document your construction standards in NAHB's *Production Manual Template* available through the NAHB Home Builder Bookstore.

Standards should include measurable performance criteria and allowable variations. Standards should be written in a way that is easy to understand. For example, standards should not be written using special engineering terms, but in terms understood by trade contractors and residential construction superintendents. Information on industry tolerances can be found in the NAHB publication *Residential Construction Performance Guidelines—For Professional Builders and Remodelers*. This book was produced by the Remodelors Council and Single Family Small Volume Builders Committee of NAHB.

The standards should be a major part of the written trade contractor agreements so that mutual expectation levels will be properly set for the trade and builder. The scope of work section in the trade contractor agreements should include quality control standard definitions. This section also should state that quality inspections with the builder are a part of the trade's primary duties and must be completed prior to the end of the construction phase.

In setting quality standards, we often fall into the "I know it when I see it" syndrome. Builders should include in their checklist enough detail for a wide range of professionals to use the standards as a tool. Construction managers, due to their personal experiences, are often expert in only one phase of construction. This can cause problems if the manager overlooks details in the areas that are not his or her specialty or that are taken for granted within their area of expertise. Using a complete, convenient checklist can prevent such problems.

Measuring Quality

Appendix C contains sample checklists, which are being used successfully by builders (see also Figure 6.2, which shows a portion of a filled in checklist showing how it's done). The checklists include a grading system that will be explained in the next section. The builder who utilizes the lists in Appendix C builds single-family detached housing in the moderate price range. Businesses differ in significant ways, and the quality control checklists should be tailored to the geographic tradition, type of construction, weather conditions, building codes, models produced, and other factors specific to your organization.

Checklists facilitate the management of numerous details. Keep a separate checklist for each phase of construction or for the scope of work for each

FIGURE 6.2 Sample Page from a Framing Quality Checklist

Framing Quality Checklist

	(5) 100%	(3) Mostly	(1) Some	(0) None	N/A
Wall Framing:					
1. Walls square (ck w/3-4-5 method).	☑	☐	☐	☐	☐
2. Walls plumb.	☑	☐	☐	☐	☐
3. All studs crowned consistently.	☑	☐	☐	☐	☐
4. All studs 16" o.c. or per print.	☑	☐	☐	☐	☐
5. Walls free of twisted or bowed studs.	☐	☑*	☐	☐	☐
6. All corners framed correctly.	☑	☐	☐	☐	☐
7. Corner bracing or plywood present and installed properly (nailing and 45 degree).	☑	☐	☐	☐	☐
8. Insulated sheathing stapled/nailed correctly. (Check 3 sheets add. if nec.)	☑	☐	☐	☐	☐
9. Exterior sheathing free from holes or damage.	☐	☑**	☐	☐	☐
10. Correct dimension lumber on all headers.	☑	☐	☐	☐	☐
11. Splices in top plates occur over studs.	☑	☐	☐	☐	☐
12. Wall top plate tie is evident and correct.	☑	☐	☐	☐	☐
13. Rough doorway measurements correct.	☑	☐	☐	☐	☐
14. Rough window measurements correct. (Three door measurements should be checked, additional if problems.)	☑— *Checked 3 windows*	☐	☐	☐	☐
15. Windows shimmed and plumb.	☐	☐	☑****	☐	☐
16. Doors shimmed and plumb.	☑	☐	☐	☐	☐
17. All finish nails set at doors and windows.	☐	☑	☐	☐	☐
18. Fire stop installed at cabinet soffits.	☑	☐	☐	☐	☐
19. Insulation and caulking behind tubs correct.	☑	☐	☐	☐	☐
20. Tub/shower nailers installed and correct.	☑	☐	☐	☐	☐
21. Closet drywall nailers installed and correct.	☐	☑***	☐	☐	☐
22. Medicine cabinets framed per prints.	☑	☐	☐	☐	☐
23. Attic access openings framed correctly.	☑	☐	☐	☐	☐
24. Patio/sliding doors smoothly operational.	☑	☐	☐	☐	☐
25. Garage door 2×6 jambs and brick molding installed correctly.	☐	☐	☐	☐	☑ —Not yet complete
Page Totals:	19	4	1	0	1

Comments: *Two twisted studs in front right bedroom closet wall.*

**Small holes (2) in garage gable sheathing.*

***Nailer missing in MBR closet & 1@ basement stair rake.*

****Windows are generally plumb but serveral not shimmed.*

trade. To ensure that your checklists are easy to use, keep the common items in proximity. For example, keep all basement-oriented floor framing details together. That way during a site visit you can check all those details in one visit to the basement. Keep the descriptions of listed items short and measurable with words like "plumb," "level," "to code," "consistent," "measuring 6 feet or 8 inches," "per the plans," and "per the nailing patterns." Avoid vague words or phrases such as "correct," "proper," or "appropriate" unless the standard for the word or phrase is well known in the organization and by your trade contractors.

Making the checklist measurable allows the construction manager, trade contractor, and upper management to use the checklists as a true monitoring tool. They can judge accuracy and check accomplishments. The results of an inspection are easy to document and communicate using checklists.

Regular quality control inspections should take place on each job. At the completion of each phase of construction, the superintendent or construction manager and/or the crew chief of the trade contracting crew should use the quality control checklists to inspect the job before the crew leaves. Any defects usually can be fixed immediately by the crew. This saves the trade, and eventually the builder, the costs of return visits. This policy is most beneficial on scattered sites, but even in a subdivision environment it will help you work more efficiently.

Immediate attention to the defects noted on the quality control checklist also ensures that the job site is ready for the next trade contractor. As mentioned earlier, job sites prepared for the next phase of work help the following trades work within their comfort zones.

Trades can help control quality by showing the construction manager any defects left by a preceding trade and overlooked in the inspection. If you wait for a trade to point out the defects in the preceding trade's work, you should expect to pay more to your trades. The trade who brings such problems to your attention is, after all, providing you with two services, workmanship and quality control consultation.

Checklist inspections also offer training opportunities for manager and trade contractor. The checklist can be an excellent communications tool. It gives the builder and the trade a basis from which to analyze current techniques and discuss operational improvements. A useful checklist places quality measurements in two categories: complete and incomplete. It removes much of the subjectivity from judging the quality of construction work.

Upper management must actively support the quality control process to ensure consistent attention. On a regular basis, the production manager or the owner and officers of the firm should inspect construction on site, using the quality control checklists.

Monitoring Quality Performance

Visits by company owners or upper management emphasize the importance of the quality inspections to the whole team, including trade contractors. Regular quality inspections will encourage everyone to use the checklists to monitor ongoing work. Using the checklist should become a habit for trades, superintendents, and upper management.

Quality inspections by upper management can become part of an incentive compensation program for superintendents. The three areas that most construction incentive programs emphasize are time, quality, and profitability. The checklists can be used to grade the enforcement of quality standards and allow the builder to reward superintendents and construction managers on performance that is measured rather than perceived.

The quality control checklists in Appendix C include four potential grading scores for each quality item addressed. This chart allows the inspector to recognize partial standards of compliance of each item. It measures quality by establishing a ration of points earned to total applicable points. The chart can thus be used as an evaluation form with a one-number cumulative score.

Figure 6.2 shows how the quality control checklists can be graded to determine one bottom-line measurement of quality on a particular phase of a job. This number describes the total earned points as a percentage of the total possible points on the phase of construction. Once such a grading system is operating within an organization the grades acquire a meaning of their own. Managers can use them to establish personal and organizational goals. These measurable goals then make quality standards easier to attain and maintain.

Builders can use the quality control checklists with their trades (and with each level of production management) to document and control quality. These results can become part of an annual "report card" of builder and trade contractor performance. The other success ingredients, time and profitability, should be well documented by a formal schedule and tracking system (as described in Chapter 5) and the detailed job cost records each builder maintains.

Although this book does not focus on job cost accounting, one job cost procedure is worth mentioning. Many builders budget their costs prior to construction and compare the actual costs to the budget during construction. These cost comparisons point out the "variances" from budget. You can include variances in your trade contractor report card. Most variance tracking systems use variance codes. These codes should be structured to track the parties responsible for cost overruns or savings. You can derive the number of variances a trade caused during a job or a given year and how much those variances cost or saved the builder. This information can be useful when negotiating future work with trades.

The Trade Contractor's Evaluation Form

Some builders keep evaluation forms to assess trade contractors' performance (see Figure 6.3). This is especially helpful if a builder employs several trade contractors within a given trade group. The more trades a builder works with, the harder it is to evaluate each trade's performance. An evaluation form will allow an objective judgment of performance. You can measure a trade contractor's performance on a variety of criteria, including schedule adherence, cost variances, and quality control checklist results. You also can include notes on the trade's professional appearance and comments from customer satisfaction questionnaires.

Whether you are a large- or small-volume builder, you should document your trades' performance in some way. Keep the forms in a file for each trade contractor. The actual reports may be simple or complex, depending on your management needs, but noting items in writing will help you maintain some objectivity in evaluating trades' performance. Use these full-scale evaluation forms for internal management, but do not review them in their entirety with the trades. Some quality-control tools such as quality control checklists and customer questionnaires can be shared with trades. Sharing the other elements of the evaluation form may cause a trade contractor to feel that you have been watching too closely or suspiciously, and that you have singled out that trade for special attention. You should keep an evaluation form file on all trades, but keep most of the details to yourself and your management staff.

Quality as an Image

A builder's reputation for "good quality" work depends on public perception. Your public image may or may not reflect your actual performance. Take the following industry situation as an example: A certain window manufacturer has earned a reputation with the public for quality windows. This reputation was earned in part by the quality of the product and in part by the manufacturer's excellent advertising and public relations programs. Several other window manufacturers produce products of equal or superior quality. However, in the eyes of the buying public these windows are not as good as those of the first company. Builders are similarly affected by the benefits and drawbacks of this type of public perception.

FIGURE 6.3 Trade Contractor Evaluation Form

Trade Contractor Evaluation Form

Scored on a basis of 1 (lowest score) to 10 (highest score)

Date: _____

Contractor:_____

1. Starts per agreed upon schedules: . . . .   1._____

2. Scheduled duration maintained as agreed . . . .   2._____

3. Good communication with Builder/Construction Manager .   3._____

4. Adheres to agreed upon price/payment terms for scope
   of work . . . .   4._____

5. Good job site cleanliness and professional working
   conditions (radio noise, etc.) . . . .   5._____

6. Uses trash receptacles and follows drop site instructions .   6._____

7. Maintains professional appearance (all crew members) . . .   7._____

8. Prompt and efficient response to warranty service requests   8._____

9. Good relationship with customers (understands they
   are salesmen) . . . .   9._____

10. Follows the zero defects checklists . . . .   10._____

11. Maintains a safe job site and follows OSHA guidelines
    for this phase of work . . . .   11._____

12. Attends regular training and safety meetings . . . .   12._____

TOTAL SCORE . . . . Total:_____

AVERAGE SCORE (total/12) . . . . Average:_____

Evaluation Team Captain: _____

You may know builders in your market that have a "good quality" image. This image may or may not be justified. If you are competing against such builders, you must constantly strive to educate your customers to recognize the equality or superiority of your product.

Trades and employees must be trained to realize that quality is more than materials and workmanship. The perception of quality is a result of performance, customer education, labor training, job site cleanliness, and public relations. Every aspect of a builder's operation helps to mold the company's public image. Nowhere is that public image displayed more than on your job sites. Your trades, their appearance, attitudes, and work habits all contribute to your perceived public image. In the eyes of the buying public, the quality of your work equals the quality of your public relations, the organization and cleanliness of your job site, and your customer communications. If customers lose confidence in your ability to communicate and to make the building or buying process as easy as advertised, they will lose confidence in your product quality as well.

7

Customer Satisfaction

The customer keeps the builder and, thus, the trade contractor in business. Without the customer… you've heard it all before. There is no way to overestimate the importance of good customer relations before, during, and after construction. Positive customer relations create referral business. Compared with paying for advertising and broker commission rates, each referral sale could add 3% to 6% directly to a builder's bottom line. It has been said that a satisfied customer will tell 10 people about their positive experience. A dissatisfied customer will tell 100 about the negative experience.

Positive customer relations also save in other ways. When a builder's customers have confidence, they are less likely to "create" service opportunities and/or change orders. With open and clear customer communication, most builders can avoid forcing customers to use their lawyers to resolve issues. If given the chance, the customer can actually become a member of the builder's team.

Home buyers are making one of the largest investments of their lifetimes. They are excited and thrive on seeing and talking about their new home. Happy customers spread good public relations for the builder, especially during construction. Dissatisfied customers can ruin a builder's reputation.

Because customer satisfaction is so vital, the obvious question is how do you develop it? The basic ingredient is COMMUNICATION. A buyer who feels well informed before, during, and after construction is most often a happy customer. When people feel uninformed, they are uncertain. Uncertainty breeds caution, fear, and anxiety. Their uncertainty, coupled with the level of interest they have in an investment as large as a new home, magnifies the anxiety. Home buyers may perceive a failure to communicate as the builder's attempt to cover up a problem or as a sign of a dishonest or incompetent business. When people become anxious, they react defensively. To avoid problems that can arise with anxiety, communicate openly and regularly with your customers.

Customer Satisfaction Formula

When asked what satisfies customers, most people will say a quality product, delivered in a reasonable time, and with good before-, during-, and after-the-sale service. This is 50% of the formula. The other 50% is the customer's perception of being in control and involved in the process. They want you to listen to their needs and concerns and give them an opportunity to be informed.

Today, people view a home purchase far differently than almost any other investment. They take it personally. The house that you are building will be their castle. To feel in control, they have to be part of the planning and building process. They may have questions, often inspired by the ever-increasing number of home improvement shows on television, which you and your trades will be forced to address.

Every builder who has been in the industry for a while realizes that the customer cannot run the job during construction. However, the customer will feel in control if they are aware of what is happening, why and how things are going, and if the builder respects them enough to "consult" with them often. Communicate at least twice per week with customers and allow them to feel that their input into the process is important. A scheduled meeting is the most professional type of job site visit.

The Trade Contractor's Role

Customers do not see an organizational difference among the builder's employees, trade contractors, and even outside sales agents. In the customer's eyes, all of these people represent the builder and his operation. Remember that perception is truth. Taken at face value, this statement says that the trades working on your job sites *are* your company in the eyes of the consuming public. Successful builders train employees to respect their customers and to work toward complete customer satisfaction. How many of us work with our trade contractors with this same goal in mind? Not many. More often, we make self-serving rules about no customer visits to the job site during the construction workday—not a great way to make customers feel like they are part of the team. (We wonder why they don't trust us.)

Instead of hiding your trade contractors from customers, the trades should be trained so that they are good representatives of both their organization and the builder's. They should appreciate that customer satisfaction allows success for both the trade and the builder. Every referral sale represents increased profits to the builder. This allows the builder, over time, to maintain competitive pricing and still reward the trade contractor for positive performance. Remember that mutual success is the goal.

How can trade contractors create a positive public image for a builder/ general contractor? There are several fundamental rules:

- Professional appearance—Each and every trade contractor should realize that appearance in today's marketplace plays an important role in creating the positive public image. This should be part of every negotiation and performance evaluation communication. Uniforms are the ideal but are not often practical. Possibly, in the future, the industry will move toward uniforms as a requirement for residential trade contractors. In the interim, common sense rules should prevail. Eliminate t-shirts or baseball caps with offensive slogans or remarks. Insist upon clean and hole-free jeans and shirts. Builders can set a good example by providing hats, shirts, etc. and providing them to regular trade contractors. Encourage trades to provide similar items or uniforms to their crew members.

    Insist that trades and their employees maintain reasonable hygienic standards. Personal cleanliness, shaves, and haircuts should be within publicly acceptable standards. Remember these people represent your company. They represent you to the public. Your success depends upon them. If you leave your success in the hands of someone who looks like they just spent the last month sleeping in the bowery, you are taking unreasonable risks. Weakness in enforcing common sense rules in this area is one reason many contractors in our industry have such an unsavory public reputation.

- Equipment and trucks—Equipment and trucks should reflect the pride the trade contractor has in the operation. They should not be disorganized, dirty, or in need of visible repair. They should reflect an image that will stimulate confidence to your customers. These same rules apply to the vehicles of the superintendents or construction managers employed by the builder.

- Customer communications—Many builders discourage trades from talking directly with customers. They fear that the trade or his employees will moonlight by doing work directly for the homeowner. Issues like this should be discussed and negotiated into the trade contractor agreements. They should not pose a barrier between the customer and the people working on the job site. Trade contractors should know the customer's name as well as the job number on all presold houses under construction. They should be encouraged to meet the customer if he comes on the job site. They should realize that encouraging words and positive comments about the builder or the work being done on the house will set the customer at ease and will encourage referral sales. Above all else, the trade contractor should not treat the customer as an obstacle to be avoided when they are on site.

- After-sale customer service—Warranty service is the issue where many builders let their reputations go down the drain. They have built a good product, in good time, and at a reasonable cost (as reasonable as they get these days), and have basically satisfied the customer. Then they blow all their hard work by allowing after-the-closing service work to go uncompleted. It happens to all builders at some point. It happens to many builders all the time. Why?

  Builders generally blame the trades, which is a cop-out. Builders haven't trained the trades properly to the importance of the post-sale customer satisfaction. These are the people who will refer new customers. Treat them well! The builders haven't included clauses in their trade contractor agreements addressing timeliness of warranty work. They haven't reached mutual agreement with the trade on what is warranted in the scope of work performed. The builders haven't convinced the trade that total customer satisfaction is a responsibility, not an after-thought. Often, this is because the builders haven't recognized the importance of these elements within their operation. Builders and trades sometimes take the attitude; "I'm so good. Aren't these customers lucky to have me working on their job." Don't look now guys, but there are tons of competitors waiting for you to fail.

Trades should be given a definitive procedure and timetable for handling customer service calls. Work orders, which define the request, and list the customer's name, address, and telephone number, should be issued by the builder. Additional information can include best times to call, best time or days of the week to schedule, and most importantly, a deadline date for completion.

Figure 7.1 shows a work order issued to a trade contractor by a builder. Such work orders create a sense of professionalism. They can be computerized and the resultant database of information can be used to track and report on the number of service requests for each trade, the average time expired before the work is completed, etc. All of this infor-mation can be used to track trade contractor performance. This infor-mation can be included in the trade's evaluation form as detailed in Chapter 6.

■ After-sale customer service requests should be handled in a timely fashion. The builder should have a system to acknowledge the cus-tomer's service request and to inform the trade contractor to schedule a visit to the home under warranty. The builder should negotiate into his trade contractor agreements a time frame within which the trade must schedule the work order to be completed.

I recommend a system that requires the trade to contact the cus-tomer within 48 hours of receipt of the work order. The recommended system would further include a requirement to schedule the work to be done within 10 working days of communication with owner. Obviously, these are not emergency warranty requests. In emergen-cies, the customer should be informed to call the plumber, HVAC, or electrician directly. The 10-day requirement could be waived by mutual agreement of the customer and the trade. However, a visit should be scheduled to prevent the "it dropped through the cracks" syndrome.

These fundamental rules—prompt and complete customer service, pro-fessional appearance of employees and equipment, and comfort with cus-

FIGURE 7.1 Trade Contractor Warranty Work Order

Trade Contractor Warranty Work Order

TO: CABINETS, INC. — C002   Work Order No. 1197
11597 EAST ROAD   Date: **01/18/2000**
ANYTOWN, OH 44600   Completion Required By:
                                                           02/05/2000
ATTN: Joe

You are requested to perform the work listed below
Payment shall be made by WARRANTY.

LOCATION: Ms. Blanche Griffith (07298)
1220 West Street
Othertown, OH 45300
                                                  Lot:

Home phone: (216) 123-4567   Office phone:

SUMMARY: Drawer off track, Bin rubber ring

LOCATION	DESCRIPTION
Kitchen	Cabinet drawer not on track
Kitchen	Rubber ring missing from top of trash bin
COMMENTS	Joe: Please contact owner within 48 hours to set up inspection/repair. Upon completion of work, return paperwork to our office. Thank you, Mr. Jankowski.

Work completed on:_____ by_____

Homeowner's Signature:_____

Comments: _____

_____

_____

Contractor's Name, Address, and Telephone Number
ORIGINAL – RETURN TO OFFICE

tomer–homeowner communication—are the trade's responsibility to the customer. You and your trades should remember that total customer satisfaction is in everyone's job description. Builders may fail to follow these fundamental rules of customer satisfaction because they have not recognized the importance of these elements within their operation.

Customer Satisfaction Questionnaires

After the sale, many builders ask the homeowners to fill out a survey or customer satisfaction questionnaire. These questionnaires are often given out twice, once at or shortly after closing and another several months later. The first questionnaire gives the builder the opportunity to measure customer satisfaction with the sales and construction process. The second questionnaire evaluates the warranty or service period. Both questionnaires should solicit comments about your labor quality and the attitudes of trade contractors and employees. The second questionnaire should solicit information about the customer's satisfaction with the after-sale warranty service.

You should maintain pertinent questionnaires, both positive and negative, in a trade contractor "report card" file. The information from these questionnaires enables you to make operational improvements involving your business and the work and attitudes of your trade contractors.

8

Future Strategies

Most builders find it difficult to plan long-term goals for relations with their trade contractors. Relationships with trades are changeable, and sometimes volatile. This summary chapter describes how you can tie the win/win attitude, the comfort zone concept, and an increased emphasis on training and teamwork into a basis for planning long-term business relationships with your trades. We will also take a theoretical view of the future of builder/trade contractor relationships.

The 80-20 Rule

According to the 80-20 rule, 80 percent of a builder's existing trade contractor base will perform a job to the builder's satisfaction or have the potential to do so with adequate training. This 80 percent is worthy of keeping and developing. The other 20 percent of a builder's trade base usually lacks one or more of the ingredients necessary for success. The missing ingredient may be agreeable price, acceptable quality of work, good customer service, ability to handle the volume, dependability, integrity, a positive attitude, or any combination of these. Builders sometimes continue to work with these trades, using them as backup to the more reliable trades, or because the builder has been too busy to provide training or recruit better replacements.

In most cases, the 20 percent that underperforms should be let go and replaced by trainable trades with the potential to increase the percentage of good trades in the trade base. Several years of practice with this 80-20 turnover rule will yield a vastly superior trade base for most builders. However, applying this rule involves more than simply firing and hiring trades. You should apply the concepts expressed in Chapters 1 through 7 and you must constantly try to improve the relationship with your existing trades before you replace them.

Mutual Growth—The Result of Partnering

As relationships with their trades improve, builders in a good economic climate often can expand their volume of construction. For your growth plans to succeed, the trades must be prepared to grow along with you. It is easier to help a qualified existing trade add other crew members to handle your increased volume than to find a second qualified trade to handle the extra work.

Taking on new or additional trades can threaten your existing trades who may begin to doubt your commitment to teamwork. This kind of psychological damage is hard to repair. A good trade may lose commitment to your jobs. There are times when an existing trade cannot handle the volume of work and additional or substitute labor must be used. When this happens, you should make sure the original trade understands the need for extra help by keeping the trade well informed about your production plan. Consistent communication is the key to keeping strong relationships with all your trades.

Quality trades should be included in your plan for expansion. They should be given an opportunity and assistance in developing a plan that will allow them to grow along with your business. Should a trade be unable or unwilling to grow, that trade will better understand your need for additional trades in the same trade. Compatible ambitions are another ingredient in the long-term success formula for builders and trades.

Some quality trades who wish to expand their business may need management assistance. You can make this another opportunity to strengthen the team concept. Trades often need assistance in budgeting, labor recruitment, and production capacity planning. As a growing builder, you can help the trade learn when to bring on a second crew, how to add personnel, and how to reorganize the existing crew to maintain quality standards and operate with more than one level of management. These management techniques are often foreign subjects to smaller trade contractors. Builders who have developed resources to manage their own business growth can facilitate compatible growth among their trades. Often, both builder and trade benefit from the new ideas this interchange can generate. Some builders and trades form literal partnerships with written agreements. Such partnerships can offer builders the loyalty of a dependable trade and increased authority over the trade's employees. In return, the trade has greater financial security and management support. For such a partnership to last, a builder must be considerate of the trade's other obligations. A long-term partnership can help both the builder and the trade to grow.

Two-Way Evaluation Reports

Mutual growth and teamwork depend upon open and frequent two-way communication. You should schedule regular meetings or reporting sessions in which you rate the trade's performance since the previous session and the trade rates your company's performance. If you do not have time to meet with each trade, then meet only with the major trades, such as framing, mechanical, and drywall. This report or meeting should occur a minimum of once a year, preferably two or three times per year. You and your trade should agree upon a format or agenda for these "report card" meetings. Trades can rate builders on the following:

- readiness of the job for specific trade,

- job site conditions,

- adherence to payment schedules,

- channels of communication,

- adequate lead times on labor scheduling or other events,

- adequate quality material,

- completed plans, and

- adequate lead time.

Builders can rate trades performance on the following:

- dependability,

- adherence to quality standards,

- adherence to scheduled construction start dates,

- adherence to schedule durations,

- budget variances in trades' trade areas, both labor and materials (because cross checking material-related variances, even with your trades who provide both material and labor as part of their contract, can help the trade cut costs),

- communications,

- warranty response,

- professional appearance, and

- customer relations and attitudes.

A meaningful exchange of information and perceptions will allow you and your trade to improve as individual entities and as a team. The exchange should not degenerate into a "gripe" session. Good preparation can prevent most conflicts. These meetings should involve trades with whom you have developed a good relationship, trades who have become comfortable working with your win/win philosophy. The format for your evaluation reports should be consistent for both trade and builder, creating an atmosphere of fairness for both builder and trade. You and the trade should approach these sessions with positive mutual expectations and with open minds.

Continual Negotiations

Regular meetings to exchange information allow builders and trades to negotiate on a continual basis. You can discuss problems, solutions, scope of work, terms, and price in open and nonthreatening environments.

An atmosphere of continual negotiation allows builders and trades to avoid the tension of annual "you versus me" type negotiations. This saves time and emotional stress and generally allows conflicts to be resolved outside formal negotiation settings. Even price changes can be handled with fewer surprises.

Consistent communication helps trades and builders develop successful relationships. Team members operate in an atmosphere of professional courtesy and profit potential. Once an air of optimism and respect has been established, it becomes easy to cultivate the win/win attitude in most builder–trade communications.

Communication takes on many different appearances in a relationship between a builder and trade contractor. Trade contract agreements, specifications, well-detailed construction drawings, preconstruction meetings, post-construction reviews, growth planning sessions, quality control checklists, and many other communication tools become routine opportunities to improve trade contractor relations and mutual profitability.

In Summary

Builders can take a positive approach to trade contractor relationships. You will find that not all trade contractors are willing to give the concepts explained within these chapters a chance. With patience, however, you can assemble a trade base that appreciates your efforts at developing a good relationship. Do not despair. Do not quit. Once you have such a trade base, you will find that success with this program is self-perpetuating. When you have experienced a success at building a team relationship with one key trade, you will find that its rewards both motivate you and provide a model for building relationships with other trades. Builders who work hard to establish and maintain a team will find that, as their trades' comfort zones expand, communication becomes easier and helps keep the team strong. You may also find that your own comfort zone will expand as your team begins to take shape.

The Future

The future of our industry will hold many changes. Many will either impact or be a direct result of the continuing labor shortage. The National Association of Home Builders recognizes this labor shortage and has initiated work groups to help stimulate young people to gain interest and training in residential and light construction trades.

Even the best recruiting efforts will probably not be enough. The industry will find it necessary to move toward more efficient means of constructing the nation's housing. Stick building will change dramatically. Technology will replace skilled manual labor at ever-increasing rates.

The following is a list of technologies that will gain wider acceptance as a diminishing labor force slows our production efficiencies and raises the prices of traditionally built homes and structures. This list is certainly not all inclusive and represents but a small portion of the development and research currently underway:

- Modular housing and components will become an increasing popular alternative for builders. These will include modular rooms, such as baths and kitchens.

■ Precast concrete foundations will replace poured in place and concrete block laid basements and crawl spaces. The labor timesaving will be dramatic.

■ Controlled environment (factory) based construction components will include basic wall panels, structural insulated panels, and ceiling and floor panels. These panels can be made with and without preinstalled mechanical conduits.

■ Box cabinets will become standard in even the most expensive homes and the detail and quality (and cost) of these cabinets will increase and improve.

Just as sheetrock replaced plaster and EIFS replaced traditional stucco, additional technologies will be used to eliminate the impact of skilled, on-site labor. The current wave of technology is generally expensive and at times cost prohibitive. In the near future the savings will come about as construction cycle times are shortened, labor becomes more expensive and difficult to obtain, and technology usage will drive the cost lower.

With these systems just on the horizon, now is probably a good time to start a book on how to deal with your manufacturing plant and distribution system. Good Luck!

A

Sample Trade Contractor Agreement

Warning

The sample agreement contained in this appendix is simply an illustration. It should not be used without the approval of an attorney experienced in construction contract law. Applicable law differs widely among the states, and in many cases local municipal law requires special treatment. Some of the provisions may not even apply to a particular construction project, and in such a case they might weaken a builder's position with respect to that contract.

This information is intended only to identify certain issues and highlight some alternative possibilities. It is offered as a service to builders to help them identify certain potential problems that need to be considered carefully and discussed with an attorney.

Builders should have their attorneys prepare documents that meet their particular needs.

Trade Contractor Agreement
Builder's Company Name

General Terms and Conditions

1. Trade contractor is deemed to know all of the requirements of the applicable building code for his work at the site and trade contractors' bid price is deemed to include all work and materials required in order to comply with **all** applicable building codes.

2. Trade contractor should carefully examine the site as well as any plans, specifications, quality standards and checklists, and selection sheets in order to fully inform contractor as to the existing conditions and scope of work for the job in question.

3. Trade contractor is to furnish all necessary equipment, labor, and material (if required in scope of work) to properly complete this phase of the job, and such items are deemed included in this bid price.

4. All work of the trade contractor will be performed in a good and workmanlike manner in accordance with plans and specification. Work must pass all applicable inspections, federal, state, or local, and meet any building code requirements.

5. This agreement is not offered as an exclusive contract for any phase of construction on this or any other Builder Homes, Inc project.

6. Anything which is not shown on the plans, but which is mentioned in the specifications, or anything that is not expressly set forth in either, but which is implied, and is required to carry out the work intended by the plans and specifications, shall be performed the same as though specifically mentioned.

7. Purchase orders shall set forth all work to be performed and the total contract price. All work, term, covenants, and conditions specified on the purchase order are included in the contract price. No extras or changes will be paid if not previously approved and written on a variance purchase order.

8. All work and material are subject to the satisfaction and acceptance of Builder's Company Name. Trade contractors shall submit a completed and signed quality checklist certifying completion of his work along with his request for payment. Payments made to trade contractor for work will not discharge the obligation of trade contractor for satisfactory materials or work, including but not limited to when defects are apparent at a later date.

9. Once trade contractor has started work on any job, if job sets idle with no activity for a 48 hour period, Builder's Company Name may, at their option, have work completed by others and the expense shall be deducted from the amount owed to the trade contractor or backcharged to trade contractor.

10. Trade contractor shall notify the Builder a minimum of 24 hours in advance of any material damaged or material shortage, and of any deficiencies of prior work that would cause a delay or an inferior job in his phase of work.

11. Once the work has started, the Builder assigned has the power to direct all the work from start to finish. Should there be any dispute regarding details, workmanship, discrepancies, etc., the matter will be settled immediately by the Builder's decision, and his decision shall be final.

12. After the trade contractor completes his work, he will be responsible for inspecting his work, completing the quality checklist where applicable and correcting any deficiencies before requesting payment. The Builder will not approve the invoice for payment until this inspection and completion of all items has been performed.

13. Trade contractor is responsible for any damages caused by contractor's work force, and by trade contractor's suppliers or materials.

14. Extras not billed within 30 days cannot be honored.

15. Trade contractor is to leave premises clean (house to be broom swept) and all excess material neatly stacked in garage or where directed by the Builder. Waste material, boxes, etc., are to be placed as directed by the Builder. Trade contractor is responsible for keeping any debris away from adjoining properties.

16. Trade contractor must protect all public and private improvements including but not limited to telephone, utilities, sewer, gas, curb, drive, sidewalk, phone lines, and trade contractor will be responsible for any and all damages.

17. The trade contractor agrees to pay all of his bills when due for materials and labor used on the contractor's jobs. Trade contractor agrees to keep such jobs free and clear of mechanics' claims or liens. The contractor may, at his option but is not required, make any checks payable jointly to the trade contractor and any one or more of his suppliers or workers.

18. Trade contractor must notify Builder's Company Name of any additional work performed at owners' request prior to agreeing to perform the work. Trade contractor will be totally responsible for payment by owner and warranty for any said work.

19. Trade contractor or supplier shall furnish to Builder's Company Name, copies of all specifications, installation instructions, and warranty information for materials included in their bids.

20. Trade contractor is to return to Builder's Company Name all blueprints or detailed drawings given to them for work or bid purposes. No payment will be made until blueprints and drawings are returned.

21. Trade contractor is to provide verification of current certificate of liability insurance and workman's compensation insurance and send updated copies as they renew.

22. Trade contractor is to provide a minimum of _____ ($300,000) liability insurance coverage to protect Builder's Company Name against any lawsuit that may arise as a result of trade contractor work and performance or non-performance, including material and labor.

23. Trade contractor shall be responsible for compliance with the Federal *Occupational Safety Hazard Acts* (OSHA) and all other federal, state, municipal, and local occupational hazard and safety acts, rules and regulations. Worker safety is paramount.

24. The use of drugs or alcohol is strictly prohibited on or near any job site.

25. Theft by trade contractor or any employee of trade contractor will be grounds for immediate dismissal and will be punishable by law.

26. Trade contractors and their employees are expected to be professionals. Their dress should be appropriate to the type of work they are doing and the time of year in which they are doing it. Their dress should be modest and non-offensive. Radios and noise levels should be kept at a reasonable level. They should make every effort to maintain good relationships with neighbors.

27. Trade contractor agrees to provide a one year warranty from the date Builder's Company Name closes with the owner as follows:

Trade Contractor Warranty

All work will be completed in a good, neat and professional workmanlike manner in accordance with plans and specifications and in compliance with all federal, state, and local requirements. All materials used shall be new and of good quality. All labor and material will be guaranteed against defect.

Trade contractor agrees to contact homeowner within 48 hours after receiving a request for warranty work to be performed and promptly correct the problem within 15 working days at trade contractor's expense.

If work is not completed in a good and workmanlike manner or in accordance with the plans and specifications, or work does not pass all applicable inspections, federal, state, and local, or meet all building code requirements, before payment has been made for the work, it is agreed that Builder's Company Name may withhold all payments until all problems are remedied or backcharge the trade contractor for repair or replacement. Payments on other contracts in progress may also be withheld if prompt attention is not given to the foregoing, discovered within one year, on a previous job.

If work is not completed in a good and workmanlike manner, and in accordance with plans and specifications, or defective materials are used by trade contractor, and trade contractor has quit the job and/or is no longer providing services to Builder's Company Name, a request for warranty work shall still be performed by contacting the homeowner within 48 hours after request for warranty work and promptly correcting the problem within ten working days at trade contractor's expense. However, if trade contractor refuses or does not so perform within said time periods, Builder's Company Name may, at its option, contract for other trade contractor's services to perform and correct the problem, and bill trade contractor for said expenses including but not limited to the labor and materials of another trade contractor. Trade contractor shall be obligated to pay any such bill immediately upon its receipt.

Date	Date
Builder's Company Name Representative	Trade Contractor/Company Representative

TERMS AGREEMENT—Trade Contractor

This agreement shall be in effect for a period of _____, and remain in effect until notified by either party.

Trade contractor agrees to do approximately _____ number of houses per year. Trade contractor will supply a sufficient labor force to meet the number of houses stated above. Furthermore, I/We agree that I/We will perform work on the day or days scheduled. Builder's Company Name agrees to give _____ number of days of lead time prior to day scheduled.

Trade contractor shall include as part of this agreement a *Request For Quote*, which is a list of unit prices for materials, labor, or other units commonly provided by trade contractor. All unit prices shall remain in effect *until a 30 day written notice is given of price changes*. Price changes will not be accepted on houses under contract (sold) or on issued purchase orders.

Application for payment must include a completed and signed quality checklist where applicable, signed copy of the Purchase Orders and/or Variance Purchase Order. Invoices will not be accepted, and if used, must be attached to Purchase Order or Variance Purchase Order; and of the exact amount shown on Purchase Order or Variance Purchase Order. Partial draws for partial completion of work will not be accepted unless specifically agreed upon in writing. *No* payment will be made on Purchase Orders or Variance Purchase Orders until the work is complete and is satisfactory to, and approved by, the Builder.

Payment will be made in the following manner:

_____ Net 30 days from date invoice is received.
_____ 10 to 14 days from date PO/invoice is approved.
_____ 5 to 10 days from date PO/invoice is approved.
_____ Other: _____

_____

_____   _____
              Date   Date

_____   _____
Builder's Company Name Representative   Trade Contractor/Company Representative

TERMS AGREEMENT—Supplier

This agreement shall be in effect for a period of _____, or until
_____.

Supplier agrees to make periodic deliveries to Builder's Company Name designated delivery sites as scheduled by a representative of Builder's Company Name. Supplier agrees to supply materials and or services as requested by way of Purchase Orders or Variance Purchase Orders issued by Builder's Company Name. Builder's Company Name agrees to give supplier _____ days of lead time to ensure delivery date can be met. Deliveries of material must be made in a timely manner as dictated by the Builder's and/or Trade contractor's schedule.

For payment, supplier agrees that upon complete delivery of material and/or services listed on Purchase Order or Variance Purchase Order the PO will be signed and dated. The pink copy of the signed PO is to be returned to Builder's Company Name accounting department for payment as agreed below. Under no circumstances will invoices be accepted except where accompanied by a Builder's Company Name PO or VPO incomplete and will not be eligible for payment until backordered items are received or until proper credit is issued via a Builder's Company Name Variance Purchase Order.

No Payment can be made until such time as materials are delivered to the designated delivery site in good, usable condition. Materials or supplies that are received in an unacceptable condition due to inferior quality, improper dumping or handling by supplier will be reported to the supplier ASAP. Supplier agrees to replace any materials or supplies that are unacceptable in the shortest time possible. Supplier also agrees that any cost incurred for replacement of unacceptable materials will be the sole responsibility of the supplier. The cost incurred for the delivery of backordered items will also be the responsibility of the supplier. Payment will be approved upon receipt and inspection of materials and supplies by a Builder's Company Name representative.

Payments will be made with the following terms:

A ____% cash discount will be applicable on all purchase orders paid within _____ days of receipt by Builder's Company Name.

_____   _____
Date   Date

_____   _____
Builder's Company Name Representative   Trade Contractor/Company Representative

B

Comfort Zone Checklists

As mentioned in Chapters 1 and 3, it is important for builders to recognize and take into consideration each trade contractor's needs and personal comfort zone. Comfort Zone checklists are used to ensure that an arriving trade has a chance for maximum efficiency and profit on the job site. When the items in the checklists are done before the trade arrives, the trade will be most productive, happy, and "in his/her comfort zone."

Comfort Zone
General Checklist
All Trades

Job # _____

- [ ] Prompt pay/paperwork processing

- [ ] Clear job site—ready for trade to begin

- [ ] Adequate lead/change times within scheduling process

- [ ] Inclusion in short/long term planning

- [ ] Good training
  - [ ] Use of checklists
  - [ ] Mutual inspections
  - [ ] Scope of work section from manual

- [ ] Recognition for doing well

- [ ] Attention and appreciation shown in big and small ways

- [ ] Professionalism and good leadership

Comfort Zone
Checklist for Excavators

Job # _____

☐ Detailed dig plan with grade depths and stepdowns

☐ Stairwells, chimneys, foundation changes staked

☐ Building location staked out with overdig marked

☐ Access to job site, parking for trailer

☐ Existing utility lines staked (or marked, where applicable)

☐ Natural drains marked

☐ Sewer line location or septic field marked

☐ Other, as needed for specific job

_____

_____

Comfort Zone
Checklist for Footer Crew

Job # _____

☐ A well drained work area

☐ Access and a walk path around hole

☐ Detailed prints showing all step-ups, bulkheads and level/grade changes

☐ Undisturbed or properly compacted soil to pour footers on

☐ Other, as needed for specific job

_____

_____

Comfort Zone
Checklist for Foundation Crew

Job # _____

- [ ] Access around hole, open corners, etc.

- [ ] Truss drop zone marked and out of masons way

- [ ] Square and level footings of proper width

- [ ] Gravel base in hole

- [ ] Enough overdig to ensure room to work

- [ ] Block and mortar placed in appropriate locations around and in the hole

- [ ] Detail of anchor strap/bolt location

- [ ] Temporary power and access to a sump pump

- [ ] Poured walls require additional mixer access

- [ ] Other, as needed for specific job

_____

_____

Comfort Zone
Checklist for Concrete and
Prep Flatwork Crew

Job # _____

☐ Opportunity to pour crawl space(s) prior to frame

☐ Access to site for mixers/good driveway

☐ Access area provided to basement for pour after framing

☐ Masonry ledges in garage

☐ A dry basement area

☐ No material stocked in poured area

☐ One day with no other trades on site

☐ Temporary heat in winter

☐ Porch post prep complete before porches poured

☐ Other, as needed for specific job

_____

_____

Comfort Zone
Checklist for Framers

Job # _____

☐ Access to electrical power or generator

☐ Lumber dropped at proper locations, stacked and located first in/on, first used

☐ Trusses placed in proper location, not in way of floor/wall framing

☐ Square and plumb foundation walls of proper width. Beam pockets located and layed properly. Post pads in proper locations.

☐ Detailed plans and material usage list

☐ A dry hole

☐ Dumpster/trash site close enough for access, but not in the path of deliveries and site access

☐ Special plan details discussed prior to framing start

☐ Other, as needed for specific job

_____

_____

Comfort Zone
Checklist for Roofers

Job # _____

☐ Frame complete, including felt paper

☐ Shingles dropped close to roof access location

☐ Backfill complete

☐ Drip edge installed

☐ Ridge vent hole opening correct width

☐ Other, as needed for specific job

_____

_____

Comfort Zone
Checklist for Siding

Job # _____

☐ Plans detailing all aspects sent

☐ Backfill complete

☐ On site alone

☐ No open trenches

☐ Framing inspection complete

☐ Fireplace chase complete

☐ Masonry chimneys width marked

☐ Full brick veneer complete before siding

☐ Other, as needed for specific job

_____

_____

Comfort Zone
Checklist for HVAC

Job # _____

☐ Siding contractor not on site

☐ Dry basement area

☐ Driveway access to convenient furnace entrance location

☐ Schedule timing and communication to ensure job is ready. Three day notice crucial.

☐ HVAC chases framed in proper location and proper size

☐ Concrete finished prior to HVAC finish

☐ Scheduled with plumber on larger/special houses

☐ Furnace, appliance locations marked

☐ Roof complete

☐ Other, as needed for specific job

_____

_____

Comfort Zone
Checklist for Plumbers

Job # _____

☐ One day alone in house in schedule

☐ In before electricians

☐ Schedule with HVAC only on large, very custom house

☐ Plumbing chases framed in proper locations and proper size

☐ Dry basement area

☐ Siding not on site

☐ Roof complete

☐ Plumbing/sewer access/exit locations marked

☐ Other, as needed for specific job

_____

_____

Comfort Zone
Checklist for Electrician

Job # _____

☐ HVAC complete

☐ Plumber at least one day into process

☐ Door swings marked to allow for proper switch locations—cabinet

☐ Service box location well marked

☐ Underground service trench location marked and ready

☐ Complete before drywall stocked

☐ Appliance/locations marked

☐ Other, as needed for specific job

_____

_____

Comfort Zone
Checklist for Insulators

Job # _____

☐ In before drywall stocked

☐ All cavities open for insulation or insulated during framing stage

☐ All mechanicals complete

☐ Siding complete

☐ Scheduled after electrical inspection

☐ House clean from all mechanical debris

☐ Other, as needed for specific job

_____

_____

Comfort Zone
Checklist for Drywallers

Job # _____

☐ Drywall stocked after insulation/siding/roof complete

☐ Nailers provided at all closets, ceiling/wall intersections and tubs

☐ All doorways nailed to subfloor at plate

☐ Temporary heat ready, when needed

☐ Adequate schedule time to allow quality job

☐ Access to dumpster/trash site without blocking driveway

☐ A detailed framing inspection should be done

☐ Protective plates over all potential plumbing pipe/electrical wire nailing locations

☐ Customers need to be educated to painting process prior to drywall stage

☐ Other, as needed for specific job

_____

_____

Comfort Zone
Checklist for Cabinets

Job # _____

☐ Proper order lead time and delivery schedule

☐ Painting complete in cabinet areas

☐ Good access to front door or door closest to interior location(s) of cabinets

☐ Room to work—schedule a day for them alone, if possible

☐ Trim material not in cabinet area

☐ Cabinet drawings need to be complete two weeks prior to schedule need on site. Drawing measurements need to be accurate.

☐ Other, as needed for specific job

_____

_____

Comfort Zone
Checklist for Trim Carpenters

Job # _____

☐ A detailed framing inspection complete

☐ Rough opening for doors correct

☐ Door swings marked on floor

☐ Stud locations marked where trim required

☐ Detailed set of plans with trim detail marked

☐ A reasonable period of time alone in house before finish mechanicals

☐ Temporary heat during winter

☐ Coordination with floor covering contractors

☐ Compare doors delivered to print/PO

☐ Clean work area, empty closets, etc.

☐ Interior trash can

☐ Other, as needed for specific job

_____

_____

C

Quality Control Checklist Examples

This appendix contains two versions of quality control checklists as outlined in Chapter 6. These will give you an insight into how checklists might be structured for your operation. A simpler version of these checklists would eliminate the rating scales, indicating only whether the items have been completed.

You can try to develop such quality assurance lists with the help of trades, suppliers, construction managers, and past customers. Customers can provide information via focus groups, questionnaires, and other methods discussed in Chapter 7.

Foundation Quality Checklist

Office: _____   Superintendent: _____

Job #: _____   Date: _____

Model: _____

	(5) 100%	(3) Mostly	(1) Some	(0) None	N/A
Excavation:					
Overdig 18" to 24".	☐	☐	☐	☐	☐
Hole has open corners.	☐	☐	☐	☐	☐
Minimum 3' pathway around hole.	☐	☐	☐	☐	☐
Tapered banks in sandy soil.	☐	☐	☐	☐	☐
Topsoil removed from garage area.	☐	☐	☐	☐	☐
Footers:					
Footer corners square and pinned.	☐	☐	☐	☐	☐
Maintains 4" ledge on each side of block.	☐	☐	☐	☐	☐
Footer depth per plan or code.	☐	☐	☐	☐	☐
Post pads correct dimensions and 1" below level of floor pour.	☐	☐	☐	☐	☐
Drain tile covered with a minimum of 12" of gravel.	☐	☐	☐	☐	☐
Sump bleeders installed correctly.	☐	☐	☐	☐	☐
Natural drain installed correctly.	☐	☐	☐	☐	☐
Natural drain outlet end open.	☐	☐	☐	☐	☐
Footer surface finished rough.	☐	☐	☐	☐	☐
Page Totals:	_____	_____	_____	_____	_____

Comments: _____

_____

_____

_____

Foundation Quality Checklist

Page 2

	(5) 100%	(3) Mostly	(1) Some	(0) None	N/A
Foundation Walls:					
Coved parging at footer/clock union.	☐	☐	☐	☐	☐
First course of block laid with mortar.	☐	☐	☐	☐	☐
Block wall square (check with 3–4–5 method).	☐	☐	☐	☐	☐
Intersecting walls tied correctly.	☐	☐	☐	☐	☐
Poured cores 8' on center.	☐	☐	☐	☐	☐
All joints uniform and even.	☐	☐	☐	☐	☐
All exposed joints pointed.	☐	☐	☐	☐	☐
Parging smooth and ⅜" thick.	☐	☐	☐	☐	☐
Waterproofing complete.	☐	☐	☐	☐	☐
Parging and waterproofing lap minimum 18" around exterior wall corners.	☐	☐	☐	☐	☐
Top of foundation level.	☐	☐	☐	☐	☐
Anchor bolts or straps 6' on center.	☐	☐	☐	☐	☐
Anchor bolts or straps within 1' of each corner.	☐	☐	☐	☐	☐
Interior walls clean of mortar.	☐	☐	☐	☐	☐
Footers clean of all mortar droppings.	☐	☐	☐	☐	☐
4" ledge around garage for slab floor.	☐	☐	☐	☐	☐
Beam pocket done correctly.	☐	☐	☐	☐	☐
Miscellaneous Foundation:					
Extra or unused materials stored properly.	☐	☐	☐	☐	☐
Site free of trash and debris.	☐	☐	☐	☐	☐
Attitudes and demeanor of foundation crew.	☐	☐	☐	☐	☐

Page Totals: _____ _____ _____ _____ _____

Comments: _____

_____

_____

_____

Foundation Quality Checklist

Page 3

	(5) 100%	(3) Mostly	(1) Some	(0) None	N/A
Page Totals:					
Page 1:	___	___	___	___	___
Page 2:	___	___	___	___	___
Checklist Totals (in number of checks):	___	___	___	___	___
	✗	✗	✗	✗	✗
	5	3	1	0	5
Point Totals:	___	___	___	___	___

Points Earned: (Add points in first 3 columns) _____

Scoring:

   (A) Total possible points: _____

   (B) Total applicable points: _____

   (C) Earned points: _____

   (D) Percent score: (C/B) _____

Evaluator: _____

Superintendent: _____

Date: _____

Framing Quality Checklist

Office: _____ Superintendent: _____

Job #: _____ Date: _____

Model: _____

	(5) 100%	(3) Mostly	(1) Some	(0) None	N/A
Floor Framing:					
1. Sill Sealer Evident and Correct.	☐	☐	☐	☐	☐
2. Anchor straps/bolts installed 6' o.c.	☐	☐	☐	☐	☐
3. Anchor straps or bolts within 1' of each corner.					
4. Beam level and installed correctly.	☐	☐	☐	☐	☐
5. Beam spliced at posts and nailed correctly.	☐	☐	☐	☐	☐
6. Beam supported on metal shims or solid masonry.	☐	☐	☐	☐	☐
7. Beam to floor height minimum 6'4".	☐	☐	☐	☐	☐
8. Floor joists rest on beam minimum 1 1/2".	☐	☐	☐	☐	☐
9. Correct dimension lumber for all spans.	☐	☐	☐	☐	☐
10. All floor joists crowned up.	☐	☐	☐	☐	☐
11. Double joists under all interior walls.	☐	☐	☐	☐	☐
12. End joist/rim joist space insulatable.	☐	☐	☐	☐	☐
13. Joist hangers used as required.	☐	☐	☐	☐	☐
14. Joist bridging evident and correct.	☐	☐	☐	☐	☐
15. End joist solid blocking installed.	☐	☐	☐	☐	☐
16. Extra Blocking used under patio doors.	☐	☐	☐	☐	☐
17. Double rim/header at basement windows.	☐	☐	☐	☐	☐
18. Subfloor sheathing glued and nailed.	☐	☐	☐	☐	☐
19. Subfloor sheathing spaced 1/8" all around.	☐	☐	☐	☐	☐
20. Subfloor nailing pattern correct.	☐	☐	☐	☐	☐
21. Subfloor sheathing overhangs stairwells(s) to match finished tread nosing.	☐	☐	☐	☐	☐
22. Stair risers consistent and to code.	☐	☐	☐	☐	☐
23. Stair treads consistent and to code.	☐	☐	☐	☐	☐
24. Headroom to code (minimum 6'8") on all stairways.	☐	☐	☐	☐	☐

Page Totals: _____ _____ _____ _____ _____

Comments: _____

_____

_____

_____

Framing Quality Checklist

Page 2

	(5) 100%	(3) Mostly	(1) Some	(0) None	N/A
Wall Framing:					
1. Walls square (ck w/3–4–5 method).	☐	☐	☐	☐	☐
2. Walls plumb.	☐	☐	☐	☐	☐
3. All studs crowned consistently.	☐	☐	☐	☐	☐
4. All studs 16" o.c. or per print.	☐	☐	☐	☐	☐
5. Walls free of twisted or bowed studs.	☐	☐	☐	☐	☐
6. All corners framed correctly.	☐	☐	☐	☐	☐
7. Corner bracing or plywood present and installed properly (nailing and 45 degree).	☐	☐	☐	☐	☐
8. Insulated sheathing stapled/nailed correctly. (Check 3 sheets add. if nec.)	☐	☐	☐	☐	☐
9. Exterior sheathing free from holes or damage.	☐	☐	☐	☐	☐
10. Correct dimension lumber on all headers.	☐	☐	☐	☐	☐
11. Splices in top plates occur over studs.	☐	☐	☐	☐	☐
12. Wall top plate tie ins evident and correct.	☐	☐	☐	☐	☐
13. Rough doorway measurements correct.	☐	☐	☐	☐	☐
14. Rough window measurements correct. (Three door measurements should be checked, additional if problems).	☐	☐	☐	☐	☐
15. Windows shimmed and plumb.	☐	☐	☐	☐	☐
16. Doors shimmed and plumb.	☐	☐	☐	☐	☐
17. All finish nails set at doors and windows.	☐	☐	☐	☐	☐
18. Fire stop installed at cabinet soffits.	☐	☐	☐	☐	☐
19. Insulation and caulking behind tubs correct.	☐	☐	☐	☐	☐
20. Tub/shower nailers installed and correct.	☐	☐	☐	☐	☐
21. Closet drywall nailers installed and correct.	☐	☐	☐	☐	☐
22. Medicine cabinets framed per prints.	☐	☐	☐	☐	☐
23. Attic access openings framed correctly.	☐	☐	☐	☐	☐
24. Patio/sliding doors smoothly operational.	☐	☐	☐	☐	☐
25. Garage door 2×6 jambs and brick molding installed correctly.	☐	☐	☐	☐	☐

Page Totals: ____ ____ ____ ____ ____

Comments: _____

_____

_____

_____

Framing Quality Checklist

Page 3

	(5) 100%	(3) Mostly	(1) Some	(0) None	N/A
Roof Framing:					
1. Trusses spaced 2' o.c. or per print.	☐	☐	☐	☐	☐
2. Truss clips evident on center and exterior walls.	☐	☐	☐	☐	☐
3. Gable and truss bracing evident and correct.	☐	☐	☐	☐	☐
4. Roof sheathing installed with metal "H" clips.	☐	☐	☐	☐	☐
5. Ridge vent Sheathing open 4".	☐	☐	☐	☐	☐
6. Felt paper cut out at ridge vent.	☐	☐	☐	☐	☐
7. Roof shingles nailed not stapled.	☐	☐	☐	☐	☐
8. Roof shingles overhang roof line 1/2" at gables and 3/4" at roof overhangs.	☐	☐	☐	☐	☐
9. Roof shingles have a minimum of 4 fasteners.	☐	☐	☐	☐	☐
10. First row of shingles has tar strip at roof edge. (starter strip).	☐	☐	☐	☐	☐
11. Step flashing installed as required.	☐	☐	☐	☐	☐
12. Drip edge installed as required.	☐	☐	☐	☐	☐
13. Roof is free from waviness.	☐	☐	☐	☐	☐
Miscellaneous Framing:					
1. Extra or unused materials stored properly.	☐	☐	☐	☐	☐
2. Site free of trash and framing debris.	☐	☐	☐	☐	☐
3. Attitudes and demeanor of framing crew.	☐	☐	☐	☐	☐
4. No evident floor squeaks.	☐	☐	☐	☐	☐
5. No evident floor springiness or bounce.	☐	☐	☐	☐	☐
6. Skylights (w/a) are framed properly.	☐	☐	☐	☐	☐

Page Totals: _____ _____ _____ _____ _____

Comments: _____

_____

_____

_____

Framing Quality Checklist

Page 4

	(5) 100%	(3) Mostly	(1) Some	(0) None	N/A
Page Totals:					
Page 1:	___	___	___	___	___
Page 2:	___	___	___	___	___
Page 3:	___	___	___	___	___
Checklist Totals (in number of checks):	___	___	___	___	___
	X	*X*	*X*	*X*	*X*
	5	3	1	0	5
Point Totals:	___	___	___	___	___

Points Earned: (Add points in first 3 columns) _____

Scoring:

(A) Total possible points (76×5) _____

(B) Total applicable points: (340–N/As) _____

(C) Earned points: _____

(D) Percent score: (C/B) _____

Evaluator: _____

Supt. Initials: _____

Date: _____

Improve your home building business with these Home Builder Press books . . .

Basic Business Management: A Guide for Small-Volume Builders
By Dorn Fowler

Increase your profits by refining your business management practices. *Basic Business Management* helps you set up the policies and procedures you need to become a more effective manager, run a more professional building business, and increase your margins. *Basic Business Management* answers the "whys" and "hows" of developing and implementing systems that will enable you to increase the productivity and efficiency of your resources. Each aspect of a complete management system is addressed:

◆ Management Activities and Strategic Planning
◆ Human Resources
◆ Marketing and Sales
◆ Accounting
◆ Legal Documents and Regulations
◆ Financing
◆ Estimating and Scheduling
◆ Customer Service and Warranties

Learn how to obtain financing, control costs, ensure consistent quality standards, and more effectively manage your employees and trade contractors. You'll find out how to identify market and industry trends that will yield business opportunities. Whether you are just starting your building business or you want to refine your current business practices, *Basic Business Management* will provide you with management guidance you can use to improve your efficiency and profitability.

Estimating Home Construction Costs, Second Edition
By Jerry Householder with Emile Marchive III

Bring your houses in under budget and make your construction process more efficient by developing complete and accurate estimates. *Estimating Home Construction Costs* walks the reader step by step through the process of estimating new home construction. It discusses types of estimates and when to use them, explains how to integrate estimating into other management functions, and describes the benefits of computer estimating. Householder demonstrates how to do a quantity takeoff and prepare an estimate, and he provides forms and checklists for you to adapt for your business. A floor plan, elevation, and other drawings clarify his concepts, and conversion tables are also included.

Basic Construction Management: The Superintendent's Job
By Leon Rogers

No one person has more control over the building operations than the superintendent. The superintendent's effective management of trade contractors, in-house labor, and materials has a tremendous effect on the company's bottom line. *Basic Construction Management* helps superintendents sharpen their skills in maintaining budgets, complying with schedules, and establishing quality controls. With a special focus on developing and managing systems, *Basic Construction Management* addresses the areas superintendents need to master in order to be successful:

◆ Project start-up
◆ Construction team building
◆ Quality control and inspections
◆ Trade contractor management
◆ Cost control
◆ Working with the homeowner
◆ Scheduling
◆ Safety management

Completely revised and updated, the fourth edition of *Basic Construction Management* emphasizes how to computerize your scheduling and reporting systems. Checklists and forms make this book an essential reference. New managers can use *Basic Construction Management* as a comprehensive training tool and more experienced superintendents can brush up on the latest techniques and technologies.

To place an order or for more information, contact:

BuilderBooks.com
BOOKS THAT BUILD YOUR BUSINESS

A Service of
NAHB
NATIONAL ASSOCIATION OF HOME BUILDERS

National Association of Home Builders
1201 15th Street, NW
Washington, DC 20005-2800
(800) 223-2665
www.BuilderBooks.com